The Origin of Natural Order

An Axiomatic Theory of Biology

Series on the Foundations of Natural Science and Technology

ISSN: 2010-1961

Published:

Vol. 12 *The Origin of Natural Order: An Axiomatic Theory of Biology*
by Q. Zhao

Vol. 9 *The Timeless Approach: Frontier Perspectives in 21st Century Physics*
by D. Fiscaletti

Vol. 8 *Quantum Effects, Heavy Doping, and the Effective Mas*s
by K. P. Ghatak

Vol. 7 *Magneto Thermoelectric Power in Heavily Doped Quantized Structures*
by K. P. Ghatak

Vol. 6 *Nano-Engineering in Science and Technology: An Introduction to the World of Nano-Design*
by M. Rieth

Vol. 5 *Grasping Reality: An Interpretation-Realistic Epistemology*
by H. Lenk

Vol. 4 *What is Life? Scientific Approaches and Philosophical Positions*
by H.-P. Dürr, F.-A. Popp and W. Schommers

Vol. 3 *The Visible and the Invisible: Matter and Mind in Physics*
by W. Schommers

Vol. 2 *Symbols, Pictures and Quantum Reality: On the Theoretical Foundations of the Physical Universe*
by W. Schommers

*For further details, please visit: http://www.worldscientific.com/series/sfnst

(Continued at end of book)

Series on the Foundations of Natural Science and Technology — Vol. 12

The Origin of Natural Order

An Axiomatic Theory of Biology

Qinyi Zhao

Medical Institute of China Rehabilitation Research Centre, China

World Scientific

NEW JERSEY • LONDON • SINGAPORE • BEIJING • SHANGHAI • HONG KONG • TAIPEI • CHENNAI • TOKYO

Published by

World Scientific Publishing Co. Pte. Ltd.
5 Toh Tuck Link, Singapore 596224
USA office: 27 Warren Street, Suite 401-402, Hackensack, NJ 07601
UK office: 57 Shelton Street, Covent Garden, London WC2H 9HE

British Library Cataloguing-in-Publication Data
A catalogue record for this book is available from the British Library.

Series on the Foundations of Natural Science and Technology — Vol. 12
THE ORIGIN OF NATURAL ORDER
An Axiomatic Theory of Biology

ISBN 978-981-3209-26-8

Desk Editor: Christopher Teo

Typeset by Stallion Press
Email: enquiries@stallionpress.com

Printed in Singapore

Preface

The precise concepts, strict logic, and description of a confined system of deductions of axiomatic theories make them the gold standard of scientific pursuit. When a scientific theory becomes mature enough, it should approach an axiomatic theory. Currently there is no axiomatic theory in the field of biology and all theoretical models of biology are in substance experimental or empirical. In order to gain a comprehensive understanding of biology and science, an axiomatic theory of biology should be developed. Although historically such approaches have only produced failure, many scientists have devoted themselves to such a task.

The belief established in the past 50 years is if there is such a theory of biology, it should lie on the basis of irreversible thermodynamics, which establishes the natural origin of order. With regards to this tendency, Prigogine has already made a remarkable contribution. In China, Xuesen Qian was an advocate of such research. Related research includes dissipative structure theory, system theory, synergtics theory, chaos theory, nonlinear science, catastrophe theory and cybernetics. These approaches have shown that our understanding of science and biology should be thoroughly changed at the most basic level. However, there is so far no matured theory in existence.

Inspired by such thoughts, we have (over the past 30 years) gradually developed an axiomatic theory of biology on the basis of irreversible thermodynamics and system theory. It includes protein thermodynamic structure theory and protein folding theory, biosignal theory, signal conduction theory or neural conduction, and biosignal network theory, and indeed covers all fields of biology. Most importantly, our theory has revealed the general relationship between protein thermodynamics and biological function (corresponding to one signaling network), and it provides a new way to understand biology thermodynamically. Two famous scientific debates in biology, namely the traditional Chinese medical theory and Michurin genetics, could be settled and agree with our theory of biology in full. This book comprises a chronological summary of our approaches.

The book comprises three parts. The first part describes the basic principles of irreversible thermodynamics which describes the laws of irreversible processes and acts as a physical exaplanation for the origin of natural order. Our approaches differ from conventional theories in two major aspects. First we have proposed criteria between irreversible process and reversal process, which results in the establishment of protein thermodynamic structure theory. Second, we clarify the partition function of a complex thermodynamic system which makes it possible to handle complex thermal systems by the laws of statistics.

The second part describes system theory and system logic. It provides a scientific definition of the system concept, and on this basis, the logic foundation for system theory is proposed. The system theory is completely compatible to the theory of the origin of natural order. In fact, it is a theory for the evolution of logic.

The third part describes the axiomatic theory of biology.

Although great achievements have been achieved in modern science and our understanding of nature has been widely expanded, many problems of nature and science had not been well studied. One of these great problems of science is the origin of natural order.

The origin of natural order acts as the utmost fundamental theory of our research and scientific theory and here we briefly discuss its general meaning. The reason is quite simple for we need to give an explanation for the origin of diversified matters of nature, such as people, animal, birds, stones, mountains, rivers, tables, food, soil, water, vapor, liquid, cars, culture, scientific thought itself, etc. The ancient people have known that different matters of nature are changeable and transformable and people hope to know the real mechanism for it. In biology, the protein folding, biological evolution, and biological development are typical processes within the origin of natural order. Without it, a comprehensive understanding of biology cannot be obtained. So this book was named "The Origin of Natural Order".

Throughout history, many different schools of philosophy, such as Taoism and Confucianism in China as well as Hegelism in Germany have studied this topic and many valuable viewpoints have been proposed. In fact, ancient Chinese culture was developed alongside the debate over this topic. But this book will be limited to the physical (rather than philosophical) mechanisms governing these phenomena.

Introduction

Currently, biology is generally considered experimental science. The task of experimental science is to learn new facts and phenomena, or to test the application of physical theories to known facts of biology. Historically, an axiomatic theory of biology has not yet been developed. There are considerable differences between creating models which capture the results of experiments, as compared with an axiomatic theory, the existence of which may fundamentally alter the research methods, logic, and structure of the scientific knowledge. The transformation from models of experimental science to an axiomatic theory of science is dramatic, and it requires us to change our thinking method and to recheck our knowledge of basic science and biology.

One can only understand a theory after he/she has read and understood its contents in full. Here, it is helpful to provide a logic structure of axiomatic theory as follows and so that a reader can grasp key thoughts and logic relationships among the many concepts and matters of biology discussed. It must be pointed out that only key relationships among many concepts of biology are depicted in this figure.

By inspecting this figure, we can see that irreversible thermodynamics and system logic act as fundamental theories of biology although they also rest on physical theory and logic theory.

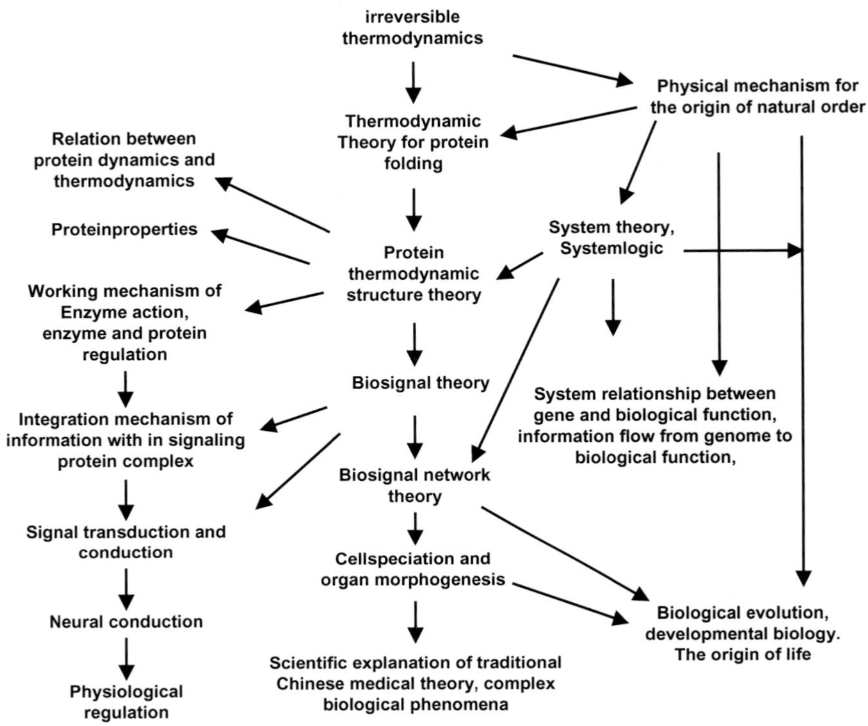

The general and fundamental relation between biological function (or network of signaling activity) and protein thermodynamics has been clearly expressed as the principle that any biological function corresponds to one thermal system within a signaling protein complex. As one biological function corresponds to thermal system within a protein complex, its behaviors can be described by the laws of equilibrium thermodynamics. This is the key thought of our theory.

There are many differences between our theory and current predominated views of biology, some of which are outlined in the following table.

Topic	Our view	Current prevailed views
Protein theory	Thermodynamic structure	Three-dimensional structure
Protein properties	Multiple thermal system	Single system of thermodynamic
Protein structure	Multiple independent spaces	Single 3D space
Integration of information	Thermodynamic mechanism	Not defined
Protein regulation	Allodynamic	Allosteric
Signal	Protein conformational signal	Not defined
Neural conduction	Protein signal	Electric signal
Gene and biological function	System relation	Linear relation
System biology	Thermal system model	Dynamic system model
Logic	System logic	Elementary logic

Acknowledgements

I would like to thank all scientists who have supported me, given suggestions and granted me courage to explore a new world of science and overcome difficulties ahead in my life. This book is also dedicated to all people who have free will in their way perusing truth in the world.

Contents

Part I

Physical Foundation

Chapter 1

The Essence of Irreversible Thermodynamics

Summary: The basic concept of irreversible thermodynamics is discussed in this chapter. We show that in all movements (processes) of irreversible thermodynamics, there is a giant space in which all changes of thermodynamics obey the principle of equilibrium thermodynamics, but their features cannot be fully explained by it. The new order can be generated in the processes of irreversible thermodynamics. Thus, the origin of natural order can be fully explained by its principle.

In view of thermodynamics, all natural changes of nature, including biological changes, are processed irreversibly. Therefore, the essence of life must be explained by principles of irreversible thermodynamics.

Philosophically speaking, the irreversible thermodynamic changes can result in chaos. Many scientists working in different fields, such as physics, mathematics, chemistry and biology, have studied this phenomenon by applying different methods and have described it using different words. Intercommunication between the different groups among them is impossible in science, except that they hold similar philosophical views.

As irreversible thermodynamics is the new field of science, it is unsurprising that many debates had arisen up and that our understanding of it differs from one person to another. The precise scientific concepts of it have not been defined, leading to confusion.

However, it is a certainty that the physical explanation of biology must be based upon principles of irreversible thermodynamics (1,2).

We will discuss some essential principles of irreversible thermodynamics at the beginning of this book and will show that a misunderstanding of biology originates from a misunderstanding of irreversible thermodynamics.

This is not a textbook on thermodynamics and we assume that the reader has some essential knowledge of conventional thermodynamics. Some aspects of irreversible thermodynamics will be discussed here because of debate about it in the scientific community. These aspects are essential for understanding physical biology.

1. Reversible Process

Based upon our experience, a thermodynamic system, when stimulated, will be subject to irreversible change.

In order to analyze the physical process of this system, the reversible (or quasi-static) process has been proposed and theorized in the history of physics.

If a stimulus acting on the thermodynamics system changes very slowly, the equilibrium of the system will not be disturbed. Physical change meeting this criterion is called a quasi-static or reversible process.

The quasi-static process is critical to thermodynamics and shows the following characteristics:

(1) The final state of the thermodynamics system is the most stable.
(2) The thermodynamics function, such as Gibbs free energy, entropy, can be calculated by the principle of equilibrium thermodynamics.

The concept of quasi-static process has been well accepted by the scientific community. In traditional thermodynamics, which studies the thermodynamic state of ideal gas, it is not difficult to judge if one physical process is quasi-static or not. However, in our study of a more complex thermodynamics system, such as protein folding or macromolecules, it is very difficult to judge it both theoretically and experimentally.

2. Irreversible Process

All changes of nature are a result of an irreversible thermodynamic process. There is no quasi-static process in nature. We consider one change of thermal motion as a quasi-static process in a specific case. In an absolute sense, they are not quasi-static processes in nature.

Although there is no quasi-static process in nature, we must study irreversible processes by concepts obtained from our study of the quasi-static process or by principles of equilibrium thermodynamics.

If a thermal system changes via way of quasi-static process, the system will be always at equilibrium state or at its most stable state under all conditions of change.

Here we will discuss several principles of irreversible thermodynamics that could be applied to many disciplines, particularly in biology.

Principle 1. For a system which is subject to change irreversibly, many sub-systems could be isolated in irreversible thermodynamic processes and they are always in an equilibrium state.

Principle 2. Any irreversible process could be divided into many quasi-static processes and mechanical processes.

These two principles are very simple and they are well compatible with conventional views of thermodynamics. These principles make us analyze irreversible processes theoretically in a simpler model.

For example, when wind blew against wood, we heard sound. The quasi-static process cannot emit sound; sound is produced in a mechanical process of the air.

A highlighted feature of the irreversible process is that we cannot calculate the final state of the thermal system. It can reach many different states via different ways.

3. Concepts of Irreversible Thermodynamics: Freedom of a Thermodynamics System, Pother and Potherse

In order to analyze the irreversible process theoretically, we must find and formulate a detailed physical parameter that could be utilized in the field.

For a complex thermodynamics system, we must analyze the relationship between different thermodynamics processes. The relationship between them plays a central role in our study of the behaviors of complex systems.

In conventional thermodynamics, there is a physical concept that defines the relationship between different physical processes. It is the half-life of a thermodynamic process.

Therefore, the relationship between different thermodynamics processes can be analyzed by this concept.

However, we cannot analyze the detailed features of the thermodynamic process based upon a half-life concept. We must use another concept: the freedom of a thermodynamic system.

For a simple thermodynamics system,

We define:

$$\mathrm{T} = \mathrm{S}/\mathrm{v}$$

Where T is a freedom of a thermodynamic system, S represents the sample space of this thermodynamic system. And v is the rate of its basic thermodynamic motion (pother).

The physical meaning of this equation is clear.

Let us consider a simple thermal system as follows. If a system has S thermodynamic states which are at same energy level, these state are independent of each other, and the transition rate between these states is v. If the system acts on mechanical stimulation, the time (T) is consumed in recovering its equilibrium thermodynamics state.

For a system with N independent pothers or particle, the sample space (S) of their thermodynamics state is N!.

Please note the following equations:

$$\mathrm{S} = \mathrm{N}!$$

$$\mathrm{T} = \mathrm{S}/\mathrm{v}$$

This estimation is rather crude and errors may occur when we calculate the T theoretically. However, when we analyze the relationship between different systems, it is rather precise.

There are many thermal motions considered for a complex thermodynamics system. We call the basic unit of thermal motion of this thermodynamics system "pother". A thermodynamics system represents a space of pothers and this thermodynamics system is dubbed "potherse".

The elementary concepts of this theory are pother and potherse which are used in our discussion of protein structure and properties.

Thus, any thermodynamics system has its particular Freedom.

4. Independent Space of Irreversible Processes in View of Thermodynamics

We can analyze the independence of irreversibility based upon thermodynamics, although we have not gone into detailed mechanism for it.

The simple analysis reveals that there is a giant space in a range of irreversible process and all changes in this space obey the thermodynamics law. Their features cannot be fully explained by the principles of thermodynamics. We refer to this space as independent space (IS).

Figure 1 illustrates the basic characteristics of irreversible process.

When a stimulating factor (S) continuously acts on a system, the system will be subject to change. ①, ②, and ③ represent three pathways of the irreversible thermodynamic process. ④ represents

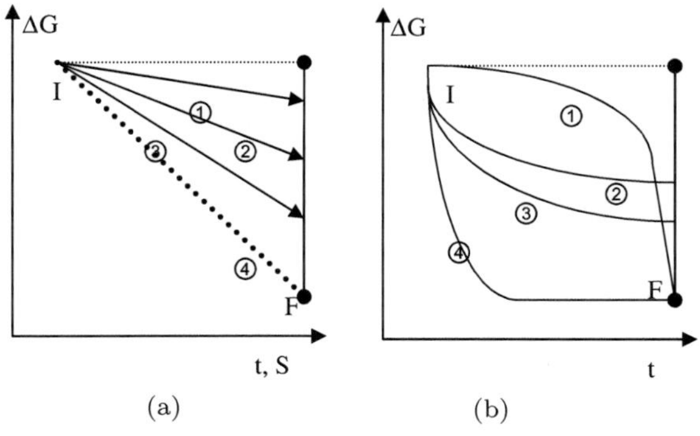

Figure 1. The independent space of the irreversible thermodynamic process.

ΔG (Gibbs free energy of the system), t (time), S (stimulation to the system), I (initiating state), F (final and equilibrium state of the system).

quasi-static (reversible) process. The system cannot reach the equilibrium state (F) by way of an irreversible process. Thus, the final state of a system is logically independent of any thermodynamic property of a system, but dependent on a pathway of system change. The property of pathway has no logical relationship to any property of the equilibrium state of the thermodynamics system. In chemistry, we know that the property of a catalyst has no relationship to any thermodynamic property of a reaction. In biochemistry, protein folding cannot be fully explained by the principles of equilibrium thermodynamics. B) When a system is suddenly stimulated, the irreversible process can lead a system to diversified states in many ways. Theoretically, some of the ways (① and ④) of irreversible processes can lead a system to its most stable state.

In Figure 1, we see that ΔG (difference in Gibbs free energy between two thermodynamic states) represents the independent space. The irreversible process can lead a thermodynamics system to any point in this section. Features of the irreversible process (pathway) determine the resulting state of a thermodynamics system. The thermodynamics function of a thermodynamics system cannot determine its final state.

In chemistry, the pathway of irreversible process cannot be attributed to any property or principle of equilibrium thermodynamics.

We all know that catalysts accelerate the rate of the chemical reaction. If many reactions can occur in the same environment, the catalyst can preferentially accelerate one direction (or reaction) without influencing other reactions. Therefore, the final state of a chemical system is determined by the catalyst.

The dynamics of a chemical reaction provides another good example. All chemists know that the activation energy of chemical reaction has nothing to do with the thermodynamic function of a thermal system. We cannot calculate it based upon thermodynamics data of materials.

The cognition of the independent space of irreversible process is of great significance in science. We will show that the new order can be generated in this space and the new matter can own its particular features that are not due to any thermodynamics properties of a system.

5. The Independent Space Becomes much Larger When the System is Enlarged

If the thermodynamics system becomes much larger, the independent space (IS) increases rapidly.

Let's take a simple calculation:

If one thermodynamics system has 100 independent particles, the collision frequency is 10^{10} turns per second. The thermodynamics sample apace of the system is 100!, then

$$T = 100!/10^{10} > 10^{90} \text{ second} > 10^{86} \text{ hours}$$

This is a very long time, and is out of the scope where we can feel it. This demonstrates that the timeliness of thermodynamic change cannot be ignored.

We picture the relationship between volume and the IS in Figure 2.

Where S represents the system; T represents time, A represents the time demanded in the irreversible thermodynamic change by which the system reaches its thermodynamic stable state from any beginning, and B represents the time needed in system expansion; The IS represents the difference between A and B.

The A is a factorial function of the space. The B is in proportion to the space. As the system becomes larger, A will increase rapidly compared with B. Then the process of irreversible thermodynamic change will be thoroughly independent of its equilibrium thermodynamic principles. Then, the process of irreversible thermodynamic processes has nothing to do with the thermodynamic property of the system.

Thus, the universe will find no way to reach the heat dead state, but will remain changeable in the IS (independent space). The heat death concept of the universe is incorrect.

This interpretation does not express that the second law of thermodynamics is wrong, but we must restrict its definition. If the

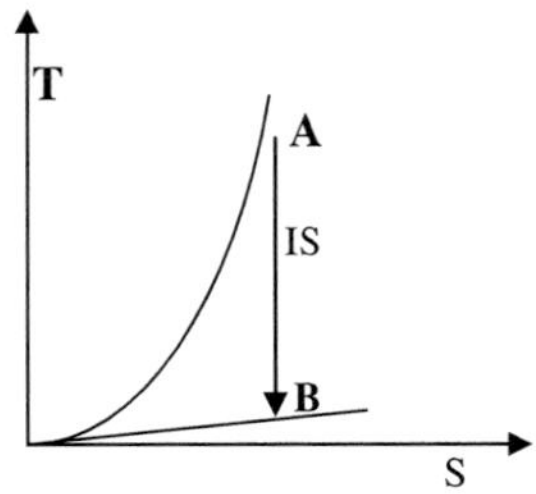

Figure 2. The differences generated between the time (B) required for the expansion of the system and time (A) taken in the transformation of the system to reach its thermodynamically stable state.

In this figure, we can easily see that when the system becomes larger and larger, the time for transformation of the system to reach its thermodynamics stable state becomes much larger. The time required for the expansion of the system is in direct proportion to the diameter of the system. The time required for the arrival of the thermodynamics equilibrium is in direct proportion to the N!, where N represents the quantity of the space. The difference (IS) between them is very large when the system becomes much larger. Thus, when the system becomes larger and larger, it will never reach its thermodynamic equilibrium. Then, the process of the irreversible thermodynamics processes will be completely independent. All the points of the IS represent the possible state of the system. The IS expresses the giant space in which all thermodynamic changes obey the equilibrium thermodynamic principles but some of their features cannot be described by the equilibrium thermodynamics theory.

scientific question is out of its definition domain, we must seek new principles of physics existing in the new fields of physics and study the questions using another scientific foundation, rather than based upon the second law of thermodynamics. In history, Schrödinger had previously realized the existence of the new physical laws without knowing the details of it.

The reader can understand it another way. As the concept of equilibrium thermodynamics has no dimension of time, it cannot be used alone in our study of the timeliness of the thermodynamic change; especially in cases where the change is controlled by some property of its process, rather than the property of its thermodynamic state (Final).

The unlimited variability of the thermodynamic change (or irreversible process) provides a suitable solution to the origin of the natural order which would not be contradicted by any physical laws.

In fact, the origin of natural order is a most common phenomenon of nature.

6. The Equilibrium State of Most Thermodynamics System cannot be Reached

We have shown that the independent space (IS) of irreversible process or the freedom of a thermodynamics system is very large. Here, we will discuss some natural phenomena.

In chemistry, we have the experience that the equilibrium state of a neutralization reaction between acid and alkali is processed rapidly and we have no difficulty in judging the final state of titration.

However, when we titrate the final reaction between ion and complex, the difficulty appears. We need some experience to judge the end of titration. It indicates that even for these simple processes, the equilibrium state thermally cannot be easily reached.

We all know that water vapor can become snowflakes under some conditions; in these cases, a simple molecule (such as water) cannot freeze into its thermodynamic stable state (or ice).

The same occurs in the condensation of oil, bitumen, and glass. In these cases, there is no final and equilibrium state. These materials cannot adapt to the most thermodynamic stable state since there is no way to direct these materials to it. The concept of a thermodynamic stable state, for these materials, is nonsense. However, they can adapt to a heat stable state in some cases.

Therefore, the equilibrium state of a thermodynamic system cannot be reached for some chemical and physical processes.

The reader may feel surprised at first for we normally handle such materials or questions by applying principles of equilibrium thermodynamics and why so many valuable results have not been obtained in the past. The answer is that the atoms of such material can be considered as same in some cases, and under this condition, the application of principles of equilibrium thermodynamics is right. But when the atoms of such material, such as protein, is numbered in protein sequence, and the protein conformation is sensitive to the sequence number of atoms, the case will be totally changed.

7. The Activation Energy — A Thermodynamics Concept Originated in Irreversible Process

Here, we study a common concept of chemistry which is the activation energy of a chemical reaction. It represents a typical example for the independence of an irreversible process and for the origin of natural order.

When we mix some chemical materials, in some cases the chemical reaction cannot spontaneously occur, unless we heat it. In principle, an energy barrier between the initial state and final state exist. Unless the materials gain such energy by some means, the reaction cannot occur.

In chemistry, the activation energy of a chemical reaction has been utilized to describe such phenomenon. It states that such a reaction has high activation energy and it cannot occur at low temperature. Theoretically, the activation energy has nothing to do with any thermodynamic properties of materials.

So, the activation energy concept arises up in irreversible process of a reaction. It provides a typical example for the origin of natural order. For any chemical reaction, we can find the activation energy for it. Thus, the origin of natural order is a phenomenon of ubiquity.

However, some chemists may fail to realize that we can express the activation energy in a common equation of thermodynamics: partition function of thermodynamics. We can symbolize the rate of chemical reaction as:

$$K_r = e^{-(\Delta E/kT)}$$

Where K_r represents the reaction constant, ΔE represents the activation energy of the chemical reaction, k is constant and T represents temperature.

It seems that this equation implies that chemical dynamics obey the same principles as that of equilibrium thermodynamics. Therefore, an incorrect belief may have gradually taken root in the minds of chemists. The principal difference between dynamics and equilibrium thermodynamics has been ignored.

The cognition that activation energy is a new thing originating from the dynamic process of a chemical reaction is of great significance. Our discussions about the origin of natural order are based upon this conclusion.

8. The Process and the Result

The above analysis results in an important conclusion: the separation of the process and the final equilibrium state of a thermodynamics system. In other words, the properties of the irreversible process of a thermodynamics system are independent of any thermodynamic properties of it.

This concept is very important for our discussion of the origin of natural order.

Equilibrium thermodynamics has no concept of time. The timeliness of thermodynamic change must be encountered by the simple application of equilibrium thermodynamic principles. In such cases, some additional hypotheses must be introduced, but these hypotheses have no meaning in equilibrium thermodynamic theory. In order to calculate the rate of chemical reactions, the activation energy concept of a chemical reaction has been introduced. However, the activation energy has no relationship to the equilibrium state of a thermodynamics system. Although we can calculate the rate of a chemical reaction, we cannot do it by equilibrium thermodynamic parameters of the reaction. Although equilibrium thermodynamic parameters are necessary in these cases, they are not handled with chemical change alone.

Thus, the equilibrium thermodynamics theory alone cannot describe all features of irreversible thermodynamic process. The chemical dynamics differs from chemical thermodynamics in principle.

Some characteristics of the irreversible thermodynamics process of nature are noted in our discussion of the process and the result of the irreversible thermodynamic processes.

Here we outlined several aspects of our discussion.

Firstly, our study of irreversible thermodynamic process by equilibrium thermodynamics focuses on the result of the irreversible thermodynamics process, rather than the properties of thermal process. Many scientific questions of our discussion have not been studied by applying equilibrium thermodynamics theory, and will not be so. When we understand the difference between result and process, all questions could be resolved easily.

Secondly, the equilibrium thermodynamics theory is only fruitful in some simple cases. If the situation is very complex, we cannot apply the theory to the actual problems.

Third and most importantly, some irreversible thermodynamic processes are not describable by the equilibrium thermodynamics theory. Therefore, the equilibrium thermodynamics theory cannot analyze this type of the irreversible thermodynamic processes. For example, where is the fend result of social change and of universal change? The transformation of glass from a heat-stable state to a thermodynamics stable state cannot be observed in practice. In most cases, we can assume that this transformation does not occur.

Fourthly, even if the result is the same in some cases, the processes of the thermodynamic system change cannot be ignored.

Finally, in most cases of irreversible thermodynamic processes, the process of the system influencing the actual state of the system determines the subsequent change of the system. This feature cannot be explained by equilibrium thermodynamics. For instance, we cannot imagine life without a second world war. Although WWII did not affect all aspects of our society, it irreversibly influenced social change, especially with regard to our present society. We have no method to compare life with and without the second world war.

Common people realized this scientific feature by their experience, which is part of our language. For example, "time and tide wait for no man".

Equilibrium thermodynamics was developed by studying the ideal gas. Once established, it was applied to other fields, such as liquid and solid. Generally, the thermodynamic system, which was studied by traditional thermodynamics, is smaller. As thermal motion of molecules of gas is very much faster, the inner equilibrium thermodynamic state of a gas could be rapidly arrived at. Normally, the change of physical stimulus is always slower than the change of inner thermodynamic.

If the case becomes very complex, the process of irreversible thermodynamics change will be independent to any property of its most stable state of a system. The order can be generated in irreversible process.

The independence of the irreversible thermodynamic change to its equilibrium thermodynamics principle makes us reconsider our scientific thought and views. It enables us to understand some questions based upon a new scientific foundation.

The independence of the process indicates there is a giant space within this space. All changes obey the second law of thermodynamics, but their features cannot be fully described by the second law of thermodynamics (see the following discussion).

9. Some Reasons for Our Conclusion

Some readers have difficulties understanding our conclusion. They may question why all chemists have not mention this independent space before.

We believe several reasons have prevented chemists and physicists to recognize this principle.

(1) In conventional physics, the theoretical subject of a thermodynamic system is the ideal gas. The IS has been eliminated by hypothesis of quasi-static process. If we study scientific problem on the basis of this concept, the IS cannot be found.
(2) In most cases, chemists can subconsciously select an experimental model and this property could be ignored within these experimental models.
(3) Chemists believe that the principles of thermodynamics can be applied anywhere and that everything could be resolved by these fundamental principles of science. Few scientists pay close attention to the invalidation of the thermodynamics principle in new fields of science, (complex systems) and the categorization of new scientific concepts.
(4) As many scientists, such as Prigogine, have claimed, conventional physics, as well as chemistry, give us no comprehensive understanding of nature, but only fragmented views of the world, even

the world of materials. However, they study dynamic system, not thermodynamic system.

(5) Most scientists are interested in experimental models that are useful to practice. The correctness of experimental models can be judged experimentally. But these models are unable to provide a theoretical explanation for many natural phenomena as well as for theoretical questions.

(6) In an ideal gas, all atoms are same and we can ignore the difference among different atoms. However, the case is fundamentally altered in biology. The amino acid residues of a protein are numbered. They are not the same nor could they be annexed by any methods. Therefore, numerous thermodynamic states could be generated in protein folding.

Engaging in a new theoretical study of science is almost equal to rechecking the validation of scientific laws under an expanded scientific sight or definition domain and explaining their relationship. The new theory will not destroy all prior knowledge of science, but will define the domain and will structure it under a new scientific framework.

Theoretically speaking, scientific revolution is not due to new discoveries, but due to our thought processes in reviewing them. Thus, in order to understand a new scientific theory, we must first question what we have ignored and reconsider the natural phenomena with which we are acquainted but have never studied scientifically.

References

1. Kondepudi D. and Prigogine I. Modern Thermodynamics: from heat engine to dissipative structure. 1998 John Wiley Press, New York.
2. Baierlein, R. 1999. Thermal Physics. Cambridge University Press. Cambridge.

Chapter 2

The Origin of Natural Order

Summary: In irreversible thermodynamics, the generation of a sub-system within a thermodynamics system represents the fundamental mechanism for the origin of natural order. According to this theory, the origin of natural order is the most common phenomenon of nature. This new understanding makes us analyze biology theoretically.

In the view of irreversible thermodynamics, all things are the product of the irreversible process of nature and one thing corresponds to one thermodynamic system.

Therefore, the question of the origin of natural order could be analyzed by irreversible thermodynamics.

The detailed mechanism of the origin of natural order varies from one thing to another. In order to gain understanding of one thing, we must study its specific properties.

In this chapter, we will study some general features of the origin of natural order.

1. The Origin of Sub-systems in a Thermodynamics System in Irreversible Process

When thermodynamics is subject to change irreversibly, many sub-systems could be isolated within it theoretically and always in the equilibrium state.

These sub-systems and their relationship represent one type of natural order and the production of these sub-systems is the detailed mechanism of natural order.

According to the features of thermodynamics systems, we discuss several typical cases:

(1) The simple system, such as gas and fluid:

When gas acts a mechanical stimulation, it can certainly be divided into many sub-systems of thermodynamics although we cannot see and isolate it physically. But these sub-systems are not stable, they are changeable with time, and finally all sub-systems disappear. Gas can emit sound, and sound represents one type of order. Although it can certainly generate some types of physical order, it has nothing to do with our study of biology. But the cognition that it can generate physical order is significant for our theoretical study.

We can judge the existence of such sub-systems by chemical tests and phenomena. When we drop sap from a flower on a white filter paper, a varied color figure can be seen on the filter paper. These figures represent order. In fact, a useful method for isolating chemical materials has been developed based upon this phenomenon, chemists know it as the chromographoresis method.

According to the conventional view, such a system is simple, and shows no ability to generate physical order. Some scientists utilize the words "linear science" to describe behavior under stimulation. We hold a different view. Although they respond to stimulation in a simple fashion, they have the ability to generate order. However, in most cases, the order generated is unstable and disappears soon.

(2) The dynamics system.

Physicists and chemists pay much attention to these systems. The sub-system can be seen by the naked eye, such as a whirlpool of water. The study of these systems generates dissipative structure theory and chaotic theory. As the method and principles adopted by physicists differs from our theory, we will not discuss it here. But we point out that all studies on dynamics systems are specific mechanisms of the origin of natural order, and are not the generic principles of the origin of natural order.

(3) The thermodynamics system can generate stable order.

These systems are of particular significance in physics and biology. The protein folding, the packing of chromosomes, the generation of biological organs are all example of this mechanism.

Water can form a snowflake; the snowflake is the production of irreversible thermodynamic process. The snowflake is heat stable, but thermodynamically unstable, when it is formed, at least theoretically it can retain its state forever.

The condensation of bitumen and plastic are also examples.

In this mechanism, the exchange between system and environment is unnecessary. The open system concept is also not required.

In physics, the atom, molecules, earth and sun, are all examples. They are heat stable, and when formed, they will exist forever. In fact, the recognition of this type of origin of natural order has led us into a new field and finally the axiomatic theory of biology has been proposed.

I Prigogine realizes that we must study the biological processes by irreversible thermodynamics and he believes that his theory will explain more properties of biology. However, the experimental scientist does not accept his approach. The key reason, I believe, is that he fails to realize this type of mechanism of the origin of natural order. Then he cannot reference real questions of biology although his philosophical views are quite correct.

2. Time and Space are Fundamental Factors that Control the Origin of the Natural Order

Being aware that the origin of natural order is equal to the production of sub-systems in a thermodynamics system, we study the condition for it.

According to physical principles, time and space are important factors that control the origin of natural order.

Any changes occur in time and space. The t provides fundamental condition for the origin of natural order.

A system, when we see it by a diversified scale of time, can show a different figure.

Time, when we discuss a single physical change, is merely a simple physical parameter. However, when we discuss some complex physical changes and there is complex relationship between them, time becomes a most important cause of some processes. In a complex system, it defines the direction of the thermodynamic change.

The common people realized this scientific feature by experience, which is recorded in our language, such as "time and tide wait" for no man.

We have discussed this in Chapter 1.

Space is another fundamental factor that can influence the irreversibility of a thermodynamics system. When a system becomes very large, its equilibrium state cannot be reached.

From equation:

$$\mathrm{T} = \mathrm{s}/\mathrm{v}$$

$$\mathrm{S} = \mathrm{N}!$$

We see that the S is the function of space.

3. There is No Thermodynamics Limit for the Origin of Natural Order

We have seen it is unnecessary to appeal to complexity for our understanding of the origin of natural order. The origin of natural order is a simple matter in view of its fundamental principles.

The fundamental mechanism of the origin of natural order has nothing to do with thermodynamics, although it is processed irreversibly in view of thermodynamics.

We can dissect irreversible processes into mechanical and equilibrium thermodynamics processes. Order is generated in mechanical process. It merely unfolds the potential stored in its thermodynamics state.

So, we gain a surprising result. We can analyze the origin of natural order without referring to the principles of thermodynamics.

This conclusion can also be reached by the separation of process and result of irreversible process.

Prigogine believes that order can be only produced from a thermodynamics system that is far away from its equilibrium state. It is correct for some cases, but cannot be applied as a rule.

4. Unlimited Changes of Systems

We have studied that the new order can be generated in the irreversible process of a thermodynamics system. The principles are rather simple:

The principles and mechanisms described in thermodynamics are rather theoretical and it is very difficult to apply these principles into practice. We had to find an easier way to do so. According to the basic principles of the origin of natural order, we promoted several criteria that could be utilized to study the natural order in a more convenient way.

(1) The unlimited change of systems provides essential space for the new thing. The unlimited change of system is the prerequisite for the origin of new rules.
(2) Without the new order, the system will become complex. The new order (principle, condition) can control the change of systems.
(3) The properties of a new order cannot be traced to any fundamental properties of systems.

Unlimited variableness exists in all systems. As the quantity of matter increases, its variableness will increase rapidly. If the new rule or the restrictions were not imposed on the system, the system would become confusion. The unlimited variableness of the system is the foundation of this property. The unlimited variableness is also the prerequisite for the origin of the natural order. If variableness is not unlimited, according to the views of mathematics, at least, it is meaningful. According to the views of mathematics, the unlimited variableness may not be meaningful. As we will discuss in following chapters, the system logic is divergence. If the restrictions or new rules have not been imposed on to restrict change, the system will become chaotic.

The uncertainty of the matter or the confusion of the system is the condition for the origin of the natural order. The confusion state of the system is the symptom for the origin of the natural order. When we meet it, we could find some way to avoid it and the new rule would be established. This is common sense of social knowledge which we all understand. But it must be pointed out that confusion is only the prerequisite for the origin of the natural order. The confusion of the system can retain its state forever.

The structure represents one type of the order. If we could find one type of structure, it must be produced by one specific mechanism of the origin of natural order.

5. Even the Order could be Generated in Mathematic Logic Change

Mathematics logic is believed to be the most precise science. However, logic confusion has been commonly observed in this field. "The certainty of the mathematics has been lost", said Morris Kline (Mathematics: The loss of certainty. Oxford University Press 1980 London).

But the certainty of mathematics has not been fully lost. Mathematics still is the most precise knowledge among all types we know. The uncertainty of mathematics comes from its unlimited change of the logic. However, when the conditions and restrictions have been imposed on mathematics logic, they cannot be transformed with each other freely. The certainty of the mathematics could be regained when certain conditions have been introduced into the case and mathematics logic can only evolve conditionally. Mathematics is not knowledge that purely consists of logic. Conditions and restrictions are also organized constituents of the mathematics.

The combination of the 0 and 1 could generate unlimited variableness; the computer could utilize 0 and 1 to describe all things of the world. The computer can work well based upon this unlimited variableness.

The unlimited change of mathematic logic will be confusing. For example, mathematics could not provide a certain solution to the following equation:

$$\begin{aligned} 1/(1+x) &= 1 - x^{67} + x^2 - x^3 + x^4 - x^5 + \cdots \\ &= 1 - 1 + 1 - 1 + 1 - 1 + 1 - 1 + 1 - 1 + \cdots \end{aligned}$$

If $x = 1$, we can obtain that:

$$1/2 = 1 - 1 + 1 - 1 + 1 - 1 + 1 - 1 + 1 - 1 + \cdots$$

We can arrange this equation as following:

$$\begin{aligned} 0 &= 1 - 1 + 1 - 1 + 1 - 1 + 1 - 1 + 1 - 1 + \cdots \\ &= (1-1) + (1-1) + (1-1) + (1-1) + (1-1) + \cdots \end{aligned}$$

Then, we can obtain that: $0 = 0.5$

We cannot accept these results in any sense. However, all these are carried out strictly according to mathematics logic. Thus, confusion occurs in mathematics, though it is well recognized as the most precise science. The uncertainty comes with mathematics.

In practice, we can easily define that these equations have not any means. We do not admit it, but we cannot avoid it theoretically.

The Gödel theory is of the validation of mathematics. However, he pointed out that in any mathematical system, if it is calculable, there exist some questions that are unanswerable.

This result demonstrates that even if in fields of mathematics, a science that most people believe to be a stricter logic system, confusion cannot be avoided.

In mathematics, there are many restrictions and conditions that limit the variableness of the mathematics logic. But these restrictions and conditions do not produce mathematics logic.

The confusion occurs in a computer. There are many logical problem programs of the computer. The computer virus emerged recently. This situation also shows the existence of logical confusion regarding computer programs.

6. The Order Concerning the Complex Relationship between Things

Confusion is the prerequisite for the origin of natural order. However, we must understand its broader sense. It cannot be understood simply and confusion can only originate in the unlimited change within the system.

In most case, the confusion comes from a complex relationship between things. As we discussed later, two things have many different relationships, if a new rule (or condition) is not applied to fields to restrict their relationship, confusion may occur.

In fact, this is the most important criterion for the origin of natural order.

For example, any language can be logically written by signs of 0 and 1. In fact, this is the computer language and computers work well based upon it.

Can people admit a 0 and 1 language into their daily life? Although it is possible theoretically, none have done it. Time displays its powerful effect here. Humans must select the language that is most convenient for them.

Confusion will not be encountered in the logic structure of a 0 and 1 language. But confusion can originate in the system of such convenience.

7. The Division of the Scientific Levels

The sequence of the origin of the natural order is the formation of a hierarchical structure and a division of the scientific levels.

When new things have been generated from a fundamental movement, we must study its properties at a new level. Thus the hierarchy emerges.

A newly originated thing of nature has its own properties and features, and can only be described by new scientific theory. It is a natural phenomenon.

For historical reasons, the scholar working in different fields of science describes the same thing by different terms. According to the theory of the origin of the natural order, they represent a similar relationship. However, our feelings or experiences aroused by these words are different.

Here we summarize some terms.

Fundamental things	**Advanced things**
Tao	Li-principle
Basal Tao	Advanced Tao
Basic	Advanced
Tactics	Strategy
Person	Society
Personality	Morality
Quantum Laws	Mechanics laws
Physical Laws	Biological Laws
Biological rules	Social rules
Specific Science	Philosophical science
Basic thermodynamic movement of protein	Advanced thermodynamics movement of protein structures
Protein	Protein complex
Protein complex	Cell structure
Cell	Biological organ
DNA	Chromosome
Biological Individual	Biological Group
Biological group	Biosphere
Molecular reaction	General physiological reaction
Molecular biosignal	Neural signal
Nervous reaction	Thinking reaction
Follower	Leader
Solider	Marshal
Experience	Knowledge
Technique	Science
Steam	Ice
Footpath	Road
Ideas	Theory

Things located in different scientific levels can be described by different scientific theories. There is a logic gap between different scientific levels. Although a common relationship exists everywhere, but there is not a fixed relationship between things at different scientific levels. We will discuss these problems in following chapters.

As our language is not developed based upon scientific logic, a confused situation occurs everywhere. In the English language, the usage of the word of fundamental, basic, general, and principle is not compatible with the scientific theory of the origin of the natural order. This also occurs in other language.

In English, the word "principle" means basic truth, general law of cause and effect, guiding rules for behavior, or a general law shown in working. The word "basic" means at the base or start, or fundamental. The word "fundamental" means basic or most important rules or principle.

According to views of the theory of the origin of the natural order, fundamental differs from principal principally. If water and food are fundamental requirements of all people, are they principal requirements of the human? For some people, it may be true; but to most people, it is not. Of course, in some cases, the meaning of fundamental equals to that of principal. Principle is the rule of special matter, rather than general rules of all matter. For instance, change is a general rule of all things, but it does not refer to any detailed characteristics of the things. It is not the principal feature of a thing, and we cannot identify substance with this feature.

Chapter 3

The Developing System and Hierarchical Structure of Things

Summary: The production of irreversible process is a hierarchical structure. If many types of hierarchical structure can be generated within a system, it is developing system. The nature and living body are developing systems.

All systems have the potential to generate order. However, only a few systems show great ability (potential) for it and they could be called developing systems. In order to form a developing system, several principles must be obeyed. We will discuss it in detail in the later chapters of this book. Here, we discuss some basic features of it.

The outcome of the origin of natural order is the hierarchical structure. The hierarchical structure formed in developing systems is rather complex.

Examples of developing systems are society, language, physiology of animal, nature, and thought, protein folding and packing of chromosomes, formation of biological organs, and biological evolution.

In this chapter, we discuss some essential features of developing systems.

- Conditions
- Formulations
  - Elements or variables
  - Axioms
  - Definition domain (condition for variable)

Figure 1. The expression of a system.

Conditions represent the prerequisite for the origin of the system. The formulations represent the detailed function of this system. The element represents the basic unit of it. The axiom represents principles and rules.

1. The Mathematical Expression of System

In order to analyze systems and developing systems, we must first define the mathematical expression of systems as follows:

Conditions are the prerequisites for the origin of natural order (Fig. 1). It must be pointed out that the condition of things is not the definition of things. The structure of thing endows its particular stability. For example, adults are very strong and can endure stronger blows.

2. Self-Organization versus Conditional-Organization

If a system contains all conditions for its organization, it is a self-organization system. Some examples are protein folding, development of individuals, society, and nature.

If the system does not contain all condition for its origin, it is a conditional-organization system. Examples are language and physical world.

For a developing system, both types of organization could be observed and both can meet the requirement of developing system.

The self-organization mechanism has an advantage in its stability and independence. However, it can only form linear relationships between diversified levels. The regulation between developing systems cannot be reached.

The conditional-organization has an advantage with its universal relationship between things of a developing system. However, it offers low stability.

Biology adopts both mechanisms for individual. The basic components of an individual, such as protein, adopts a self-organization mechanism for its high stability. For protein complex, it is involved in the regulation of physiology and adopts a conditional-organization mechanism.

3. The Stability and the Rate of Change Falls Downward in the Hierarchical Structure

A highlighted chrematistics of the origin of natural order is the decrease in the stability of a thing and the rate of its change.

We can see this feature both in physics and biology. Figure 2 illustrates this feature in the physical world.

In biology, we also see such a phenomenon. Bacteria can endure high temperatures, such as 100°C which humans and animals cannot endure. The division rate of bacteria is much high than that of human cells.

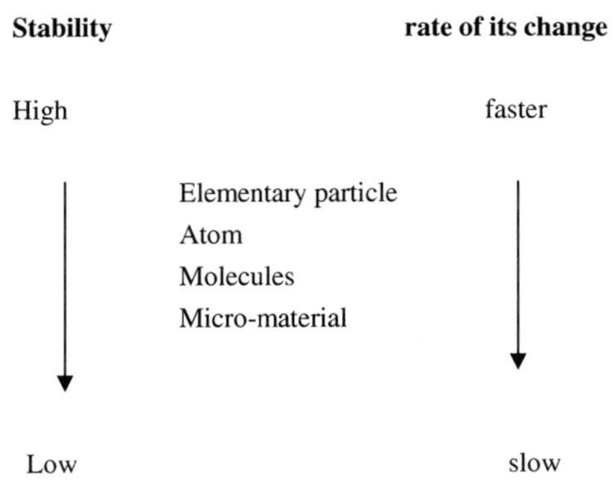

Figure 2. The change of rate and stability in hierarchical structure. The fundamental thing moves faster and has high stability.

In society, we all know that the individual is much more active than that of government. The activities of individuals are at the fundamental level for all society.

4. The Relationship between Things at Diversified Levels and the Complexity of Thing

New things show all properties of fundamental things, but possess new properties that cannot be explained by properties of fundamental things.

The relationship between things at different levels of a hierarchical structure is very complex. We cannot provide resolution for all cases, but we can obtain the following principle.

1. The unified theory that can describe all properties of a thing cannot be developed.
2. The relation of things is variable under different cases or conditions.
3. In particular case, the relation of things can be described by formal logic.

However, in views of philosophy and complexity, the origin of natural order is the process that increases the complexity of the nature. We understand it in several ways.

1. New things of nature can be produced.
2. The relationship between different things at different level has been generated.
3. Complexity has been generated. The unpredicted outcome can be generated.

5. Restrictions Applied to Fundamental Things

The movement of fundamental things is restricted. New things provide conditional factors for fundamental things. Thus, the movement of fundamental things has lost its independence in hierarchical structure of a thing.

For example, individuals are fundamental things in society. The law of society has restricted the action (or activity) of individual. A person cannot do all matters according to his will.

In biology, the movement of protein, gene expressions are under well control.

In solid and liquid, the movement of water is restricted.

The cognition of the restriction of fundamental things in a hierarchical structure plays a central role in our understanding of biology. Biology shows a typical hierarchical structure, and the advanced things, such as our thinking ability, can restrict the change of fundamental things, such as molecular reaction.

In fact, our misunderstanding of biology comes from the fact that we pay little attention to this feature.

This principle shows the theoretical limit of scientific theory on which we explain biology by our knowledge of molecular biology. Molecular biology, without the help of other theories of biology, cannot explain biology alone.

Chapter 4

Partition Function of a Thermodynamic System

Summary: The partition function is a fundamental concept of equilibrium thermodynamics. In the views of traditional thermodynamics, it encodes the particle distribution at different energy levels. However, the behaviors of the complex system of thermodynamics cannot be described by the partition function of a particle. By modifying the partition function of a particle, the partition function of a thermodynamic system has been obtained. It could be utilized to analyze behaviors of a complex thermodynamic system, such as protein conformational state, DNA conformational state. The typical fashions of distributions curves of a thermal system are of great importance and we can judge the properties of a system on the basis of it.

In traditional thermodynamics, the partition function describes statistical properties of particles (1). It is a function of temperature. As traditional thermodynamics was developed from the studying of behaviors of ideal gas, the partition function of its constituent functions of temperature has nothing to do with other environmental factors, such as volume and pressure. The pressure and other environmental factors do not affect the distribution of particles among different states or different energetic level. The consideration that energy involving in chemical reactions and physical change is normally very large and environmental factors have no great impact on the energetic level of a quantum state of atom or molecules, the traditional partition function works well in these cases.

However, the case is totally different within a complex thermodynamic system, such as protein thermodynamic state. Here we summarize some features of protein conformational state as follows. Firstly, environmental factors, such as the buffers of a protein, has great impact on protein conformational changes, and the energetic level of protein conformation was greatly influenced by environmental factor. This nature does not exist for ideal gas. Although the energetic level of a quantum state of atom, such as S or P orbit of electron, is affect by environmental factors, such as electric field, normally, the revision is very small and we can ignore such revision. Secondly, the energy involving in protein conformational change is rather small, say 1 kcal/Mol, then the revision cannot be ignored. Thirdly, the energetic levels for two protein conformations can inverse under some cases, and this feature has not been observed in quantum states of an atom.

Thus, in order to analyze the behavior of protein conformational change and properties of protein conformation (a thermodynamic system), the partition function of thermodynamic system should be developed. This chapter will discuss basic principles of a thermodynamic system and its behaviors (2). It can be regarded as the theoretical foundation in our discussion of protein conformational change in other chapters.

1. Basic Principle

According to the traditional theory, the partition function of a particle can be described as

$$N_p = \exp(-\Delta E/kT) \tag{1}$$

where N_p is the probability of a particle at the energy level E, k is the Boltzmann constant, and T is the absolute temperature.

For the ideal gas, the particle distribution at different energy levels is entirely determined by the temperature, and the energy level of a quantum state of a particle is independent of environmental factors. Since a protein conformational state represents a thermal system, environmental factors significantly contribute to its energy level.

The energy level of a conformational state can be described as

$$\Delta G^0 - F(s)$$

where G^0 represents the Gibbs free energy under standard conditions, s represents the strength of a physical variable, and F(s) represents the revision function.

Thus, the partition function of a protein conformational state, N_{ps} is defined as

$$N_{ps} = \exp(-[G^0 - F(s)]/kT) \qquad (2)$$

Under the simplest cases, the energy level of a thermal system is proportional to the strength of a stimulus. In such cases, N_{ps} can be described as follows:

$$N_{ps} = \exp(-[G^0 - cs]/kT) \qquad (3)$$

where c represents the correction coefficient.

The present data indicate that the correction coefficient of N_{ps} in the presence of denaturants, electric forces (electric fields), and volatile general anesthetics is perfectly linear. In most cases, the revision function is complex, nonlinear, and non-conforming to any known trend.

Many environmental factors can influence the same conformation. The partition function of a thermal system (or a protein conformation) can be described as

$$N_{ps} = \exp(-\left[\Delta G^0 - \sum_{I}^{N} C_i S_i\right]/kT) \qquad (4)$$

Where C_i represents the correction coefficient of the Ith factor and S_i represents the strength of Ith factor.

This equatrion has described basic principles for bio-signal or information integration within a signaling protein as we will discuss in following chapters.

Normally, a protein is a complex thermodynamic system, and the infinite variables (referring to c for the conformational states of a protein) are unknown. Thus, the behavior of a protein cannot be fully described by Eq. (4). However, the thermodynamic relation of conformational states can be demonstrated theoretically based on this concept.

There are two types of protein conformational change, namely, continuous and discontinuous. If the physical parameters (referring to c) that label a protein conformational state vary continuously or are constant, then the conformational change is continuous. If there is a sudden change of the physical parameters, the protein conformational change is discontinuous. Continuous protein conformational change does not induce alterations in the protein thermodynamics structure, whereas discontinuous protein conformational change alters it.

2. The Basic Fashion of Distribution Curves of a State of Thermodynamic System

Normally, a thermodynamic system could show many different states of thermodynamics (here, there is linguistic confusion for us normally use the same word in different case and their meaning varies). Theoretically speaking, these states represents different thermal systems which can be transformed with each other. Here system refers to actual matters (or systems) which show many thermodynamic state. The distribution curve of one state can be calculated from partition function of thermodynamic system. Here we study several typical fashion of it.

Distribution curve between two states. Protein conformational change between thermal states has been well documented in many scientific papers and textbook of biochemistry. The conclusions of our analysis agree with previous view well, but provided a unified and more detailed explanation.

The probability of one protein conformational state can be described by the following equation:

$$\mathrm{P} = \exp(-[(\Delta\mathrm{G} - \mathrm{cs})/k\mathrm{T}])/\{1 + \exp(-[(\Delta\mathrm{G} - \mathrm{cs})/k\mathrm{T}])\}$$

The distribution of the two states of a thermodynamic state is shown in Figure 1. The example of it is the open state of a channel or receptor.

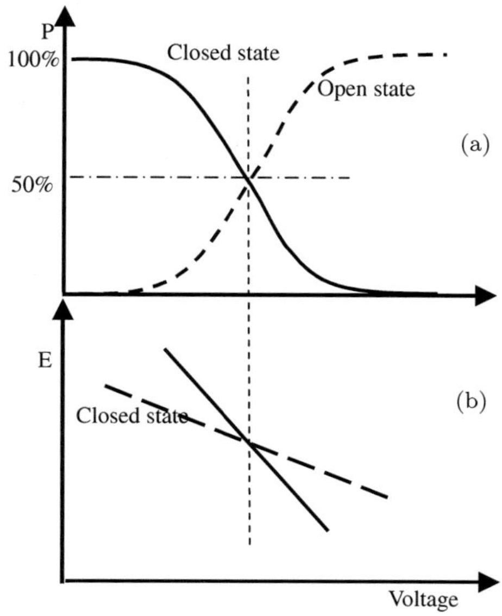

Figure 1. Diagram of distribution curve of protein conformational states.

P is the probability of protein conformational state; E represents the stability energy of protein conformational state. Figure 1A drawn distribution curves of two state of a channel. Figure 1B drawn the stability (energy) curve of two states (warning: this is not experimental data and we only drawn its general tendency in this figure). At a voltage, the stability of two states reaches the same and the probability of two states is the same. The difference of energy of the two states determines the probability of the two states.

The distribution curve of one state among three or more states. The calculation of the distribution curve of one state among three or more states is very complex. Here we will not enter the complex calculation of complex mathematic equations, but to figure our two extreme fashions of it which are shown in Figure 2.

These patterns are very common in our studying of protein thermodynamics and enzyme activity (3).

3. The Shifting of Distribution Curve Along Stimulus

The distribution curve of a thermal state describes behaviors of a thermal system in a changing environment. The shift of distribution curve could be observed under two different cases. Firstly, if the

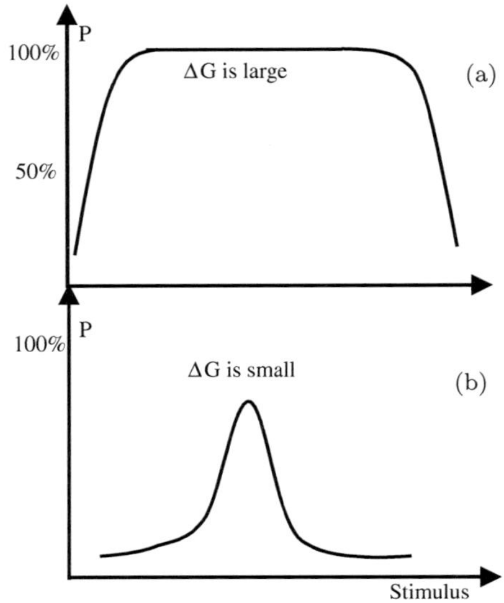

Figure 2. Two typical fashions of distribution curve of a state among many states.

The stimulus represents a value which quantifies an environmental factor. Figure 2A is distribution curve for a protein conformational state with high stability (ΔG is large), Figure 2A is distribution curve for a protein conformational state with low stability (ΔG is small).

constituents of a system, such as amino acid residues substitution, change, the properties of the thermal system will be modified and it can alter the position and shapes of distribution curve. Secondly, if the distribution curve of a system state is measured under different conditions, such as different environments or working buffer for an enzyme. If the shape change of distribution curve is ignored, it will shift along axis of environmental factors.

The shift of distribution curve of a thermal state between two states (Fig. 3) was widely seen in biochemistry, such as denaturation of DNA double helix under different salt concentration (4). Compared with the wild type of a channel, the distribution curve of the open state of a channel with gene mutation also show such phenomenon (5).

The shift of distribution curve of a state among many states is shown in Figure 4. This phenomenon is widely seen in protein

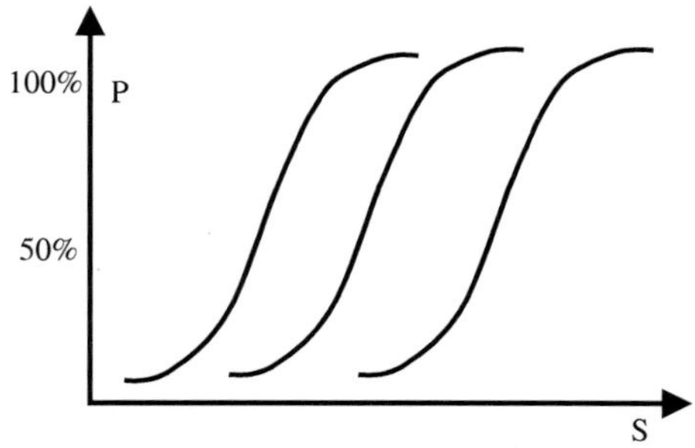

Figure 3. The shift of distribution curves along a variable.

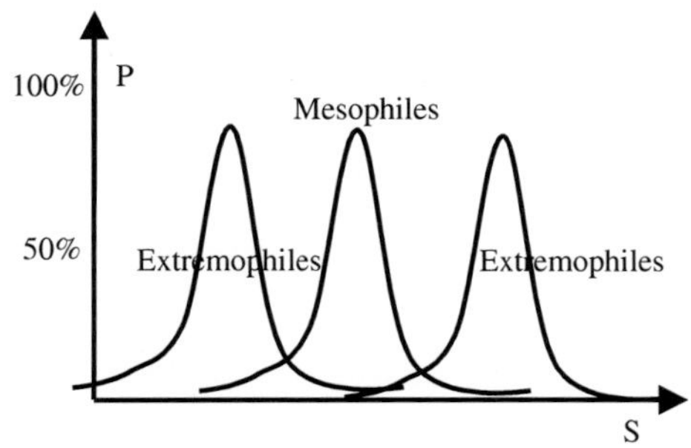

Figure 4. The shift of active conformation of a protein along temperature. Different distribution curves of active state of proteins isolated from different species which lives at different temperature is drawn in this figure.

evolution. The gene mutation of a protein can shift distribution curve of active conformation of a protein (enzyme). Via biological selection, the optimal condition of a mutant protein which is well matched up to its environment survives. Finally, the protein (enzyme) show maximum activity of a given environment that is selected.

4. Thermodynamic Oscillation of a System between Two Different States

Normally, statistics physics deals with a system which consists of numerous constituents. However, behaviors of a particle or a thermal system in a very long time can also be described by laws of statistics physics. This is true for the thermodynamic oscillation of a system between different states.

We have learnt that the energetic level of thermal system is changeable. If the energy difference between two states of a system reaches zero, a new specific phenomenon occurs. It is the thermodynamic oscillation of the thermal system. The features of thermodynamic oscillation are shown in Figure 5.

For a system which can exist in many states, many types of thermodynamic oscillation could be observed.

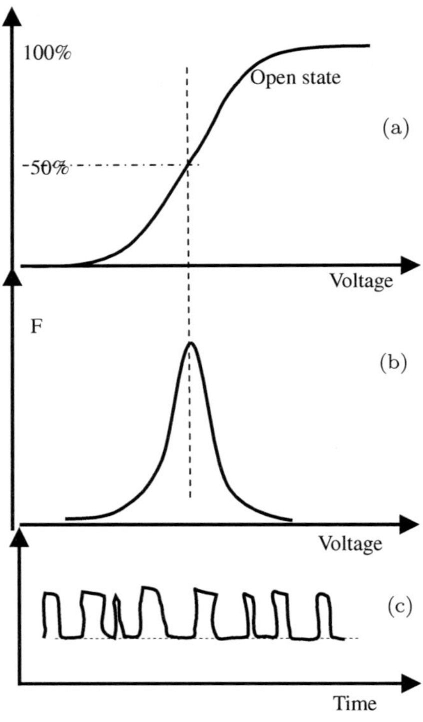

Figure 5. The principle of thermodynamic oscillation between open and closed state of a channel.

Figure 5A shows the distribution curve of open state of a channel. Figure 5B shows the frequency of thermodynamic oscillation. We can see that frequency of thermodynamic oscillation reaches its maximum value at voltage at which the energy difference between open and closed state is zero. Figure 5C is the trajectory of a channel between open and closed state.

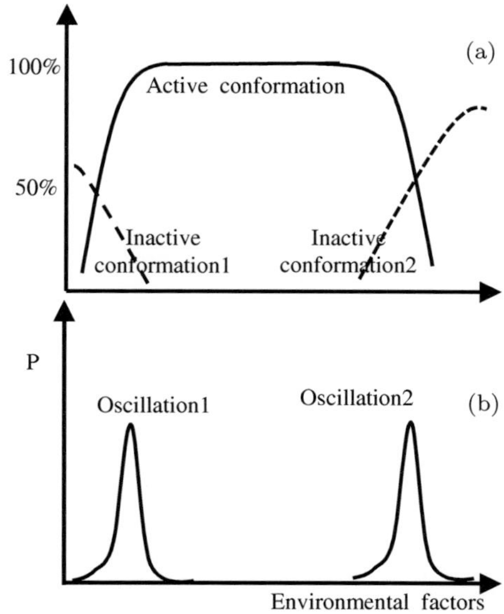

Figure 6. Many types of thermodynamic oscillation of a system.

Figure 6A is the distribution of active conformation (state) of a receptor. Figure 6B represents two types of thermodynamic oscillation: one oscillation occurs between active and inactive conformation1, another one occurs between active and inactive conformation2 (desensitization of a receptor).

For thermodynamic oscillation, the oscillation frequency is determined by many factor, such as protein flexibility, it is not the thermodynamic parameters.

The Probability ([open time]/[closed time + open time]) is a thermodynamic parameters and it can be calculated from thermodynamic parameters of the system. But in practice, it is measurable and other thermodynamic parameters of protein conformational state can be calculated from its value.

5. Multiple Dual Natures of a Thermodynamic System

As we can see in the partition function of thermodynamic system, the thermodynamic state of a thermal system is related to properties of macro-world. Thus, multiple dual natures of a system could be deduced. In protein science, a protein shows dual nature of thermodynamics and mechanics, dual natures of macroscopic and microscopic substance, and dual natures of solid and fluid.

As multiple dual natures of a protein, a protein can link up multiple properties of macroscopic and microscopic world. All information, thermal and mechanistic, microscopic and microscopic world, can be processed or integrated by a protein. This is the basic principle of biosignal theory.

General References

1. Isihara, A. (1971) Statistical Physics, Academic Press, New York.
2. Zhao, Q. (2012). Partition function of protein conformational state. *J. Comput. Theor. Nanosci.* 9, 745–751.
3. Pace, C. N. (1986). Determination and analysis of urea and guanidine hydrochloride denaturation curves. *Methods in enzymology*, 131, 266–280.
4. Schildkraut, C., & Lifson, S. (1965). Dependence of the melting temperature of DNA on salt concentration. *Biopolymers*, 3(2), 195–208.
5. Hammond, Cellular and Molecular Neurophysiology, 3rd ed., Academy Press, London.

Chapter 5

Dissipative Structure and Thermodynamic Structure Theory

Summary: In view of thermodynamics, all actual changes of nature are irreversible, and our understanding of nature, as well science, is established on the basis of principles of irreversible thermodynamics. Currently there are two different approaches in the field. First is the theory of dissipative structure which represents a dynamic but steady state of a system. Second is thermodynamic structure which represents a static state of a thermal system. The key task of this chapter is to ascertain the validation domains and different usefulness of both theories. The advantages of both theories in diversified area of science are also discussed.

Living body is a complex system of thermodynamics and the theoretical grounding of biology must be laws of thermodynamics. Without the exposition of thermodynamic principles of biological process, our understanding of biology will be empirical in substance.

Biological substances and biological processes are well organized or ordered. Therefore, a highlighting nature of biological process is the increase of order. As the second law of thermodynamics indicates that the transformation from order to disorder is general tendency for an isolated thermal system, the origin of the order in field of biology, as well as the scientific theory itself, becomes a challenging subject of science for theoretical chemists and biologists.

Erwin Schrödinger had mentioned this phenomenon and discussed it standing on the point of thermodynamics, but he had not found the answer (1-3). His approach has spurred other to study it and finally I Prigogine proposed his famous theory-dissipative structure and it provides a reasonable explanation for this problem. It states that dissipative structure could be generated in an irreversible process which is far away from equilibrium state and this type of structure could be maintained if the sustaining supply of energy was provided and material exchange between it and environment is processed. From then, there are explosive approaches based upon his theory, and which result in the establishment of chaos theory, complexity, our understanding of science has been sharpened and changed at fundamental level of science (1-3).

However, the development of experimental biology has little to do with this scientific belief. The underlying reason is that current theory of irreversible thermodynamics is not matured enough and has no ability to give satisfied explanation to numerous phenomena of biology. Therefore, new theory is needed.

In our studying of protein folding, which is widely accepted to be a typical process of the origin of natural order, we have developed a new theory for it. It says that thermodynamic structure of a matter could be generated in irreversible process and thermodynamic structure of materials is heat stable. The axiomatic theory of biology has been developed based on this cognition (4, 5).

Thus different views about this subject emerged. In order to avoid confusion of scientific community, it is necessary to carefully define the validation domain of both theories. In this paper, we will discuss the consensus and difference between two theories.

The thermodynamic structure theory of materials is completely compatible to current knowledge of molecular biology and protein science, and it takes many advantages than that of dissipative theory. It is the general principles of irreversible thermodynamics and dissipative structure that represents a specific mechanism for it. The thermodynamics structure theory could serve as theoretical foundation of theoretical biology.

1. Historic Introspection and Significance of it in Human Culture

In the beginning of our discussion, it is valuable to make a historical introspection about it for it is a fundamental question of our culture.

Although in field of thermodynamics, Erwin Schrödinger was considered the first one to study it, the origin of natural order is one of the most famous problem found in human culture. In fact, it is the main subject of the First book of China, *I Ching* (6). Many disputes about it had repeatedly occurred in China history as it serves as the theoretical foundation for Chinese culture. But Chinese thinkers did not study the thermodynamic mechanism for it, they prefer to study mechanism for the origin of social order, such as origin of rules of social and health personality. The central task of Confucian is to establish diversified social order (include personality) that are compatible with each other. In my opinion, many problems, which has been discussed in scientific community, are also discussed in Chinese history.

In essence, our understanding of nature, as well as science itself, is largely constructed based on the interpretation of mechanism of the origin of natural order. In the field of philosophy, the principle of science represents one type of order of nature. In this paper we will not discuss most of these questions, but focus on thermodynamic mechanism of it. In order to gain real understanding of it, these relationships were expressed in Fig. 1 and reader can read other papers for further studying.

In history of biology, many disputes of science arise up from different understanding of this problem. The most famous example in biology is the dispute between Michurinism and Mendelism. According to my opinion, the bio-information flow from genomes to biological function is typical process of the origin of natural order and the relationship between gene and biological function must be analyzed by system logic. The linear relationship between gene and biological function (or trait) could be isolated under a given condition among complex relationship among them.

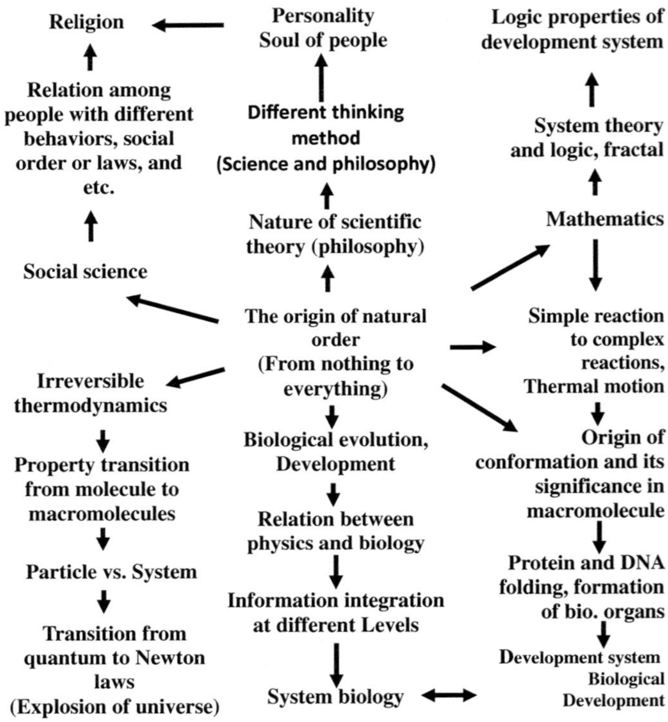

Figure 1. Diversified subjects of science related to the origin of natural order.

In this figure, we see that our understanding of science and nature is established on the exposition of the mechanism of origin of natural order. Many subjects of science expressed in this figure are far out of the scope of this book and we will not study it.

2. Differences and Consensus

Although dissipative structure and thermodynamic structure theory were developed on the different foundation of science and they differ from each other in many aspects, there are many consensuses within two theories. In order to gain comprehensive understanding of science, we have summarized consensuses and differences between two theories (see Table 1).

Table 1. Consensuses and differences between two theories.

	Dynamic system (Dissipative structure)	Thermodynamic system (Thermodynamic structure)
Sustaining Energy supply	Yes	No
Material exchange	Yes	No
Open to environment	Yes	NS
Thermodynamic system	No	Yes
Dynamic system	Yes	No
Heat stability	NS	Yes
Self-organization	Yes	Yes
Thermodynamic stability	NS	Yes
Dynamic stability	Yes	NS
Entropy controlling	Yes	No
Described by partition		
Function concept	NS	Yes
Pathway dependence	NS	Yes
Far away from equilibrium	Yes	NS
Evolutionary system	Yes	Yes
Thermal fluctuation	Yes	No
Logic trait	Complexity	simplicity

NS represents nonsense concept within a model.

The development pathways of the two approaches are also different. The dissipative theory, it was proposed by I Prigogine, was developed by chaos theory and finally stepped into none-linear science. Our studying arises up from the studying of irreversible process of thermodynamics or protein folding theory, which results in the development of protein thermodynamics structure theory, and finally system logic theory and axiomatic theory of biology were established.

3. Basic Facts and General Conclusions of an Irreversible Process

In judging or evaluating a scientific theory, we must examine basic or typical facts of one theory. Firstly, the typical fact of a theory provides a way for us to link theoretical concept and common knowledge (or our daily experience). Second, validation domain of a scientific theory could be judged by examining the range within it typical facts could be found. Third, maturity of a theory could be judged by examining features of its typical facts. The basic fact of a matured theory of science should meet two different criteria: (1) the typical fact should refer or relate to the basic process or matter of nature; (2) typical facts should be found in all areas in which a theory is validated. Then we can examine both theories by applying these criteria.

Basic facts of thermodynamic structure theory of material are summarized as follows.

The first is window frost, fern frost, and snowflake. The beautiful pattern of frost, which was produced in winter days at surface of the glass of a window, has attracted much attention of common people and scientists. It is a really enjoyments to see such things. It seems all things of the world can be generated in these figures. Standing on the point of thermodynamics, the formation of frost patterns is irreversible processes. Following conclusion have been obtained by studying it: (1) it is an irreversible thermodynamic process; (2) it is pathway-dependent; or in other words, the conditions of room influence final patterns; (3) there is a numerous space of final state (patterns); (4) the state is not thermodynamic stable, but heat stable; (5) it represents a developmental system. In this case, we have known that the lowest Gibbs free energy state of water is ice and frost is not at its most stable state, but it can be sustained forever if proper condition is provided.

Second comes from the heat treating of material, such as iron. It has been extensively studied in industry. Some conclusions have been obtained: (1) the final state is not thermodynamic stable, but heat stable; or in other words, the property of heat treated material is stable; (2) It is pathway dependent, the property of material is dependent on the condition of heat treating, such as cooling rate. (3) Different phases or structures could be generated in heat treating, such as martensite phase in steel; (4) the stability of different structure or phase differs from each other greatly; (5) at least in views of experimental science, the lowest Gibbs free energy state is nonsense and none discuss it.

Third is the asphalt. When it is heated, it become viscous, and when it cooled again, it consolidated. But it never reaches at its most stable state of thermodynamics. The lowest Gibbs energy concept is again a nonsense concept in this case.

Fourth is the chromatography technique. By this method, numerous materials can be isolated from a mixture. The process is near equilibrium and order (pure substance) is produced.

Fifth is the crystallization. The crystal state of substance represents one type of order. It is well that the crystallization could be processed reversibly. It is close to a state of equilibrium.

The aforementioned facts have served the beginning of our research. Although this knowledge is widely accepted over the scientific community and common people, its significance in theoretical science has been ignored totally before. Several reasons can account for it. First, most scientists study it in field of technology and they do not study the significance of it in view of theoretical science. Second, it is not the typical phenomena or facts for the origin of natural order and it is not suitable topic of science. Third, the thermodynamic parameter of such processes cannot be obtained from such studies and few theoretical scientists pay attention to it.

A new scientific approach often arises up from studying of problems of science without a reasonable explanation. This is particular true for the studying of dissipative structure theory. By studying complexity phenomena of chemical reaction, mainly on dynamic processes of a system, I Prigogine has successfully proposed dissipative structure theory and his theory represents predominately views in scientific community now.

The typical facts which support Prigogine theory was Benard test, Belousov–Zhabotinsky reaction, and cyclones. They were all complex and dynamic process of nature and there had not been reasonable explanation previously.

This analysis shows that thermodynamics structure theory is more mature than dissipative structure theory. It also reveals the fundamental difference between our theory and those of Prigogine's. We study simple and common phenomena of nature and simplicity, while Prigogine studies unusual phenomena and complexity. Therefore, thermodynamic structure theory is a more general and fundamental theory for the origin of natural order.

References

1. Kondepudi, D. and Prigogine, I. (1998) Modern thermodynamics: From heat engine to dissipative structure. John Wiley Press, New York.
2. Prigogine, I. and Stengers, I. (1997) The end of certainty: Time, chaos, and the new law of nature. Free Press, New York.
3. Schieve, W. C. and Allen, P. M. (1982) Self-organisation and dissipative structure: Application in the physical and social science. University of Texas Press, Austin.
4. Feng, Y., & Bodde, D. (1983). A history of Chinese philosophy (Vol. 1). Princeton University Press.
5. Zhao, Q. (2001) Irreversible thermodynamics theory for protein folding and protein thermodynamics structure. *Progress in Biochemistry and Biophysics*, 28, 429–435.
6. Zhao, Q. (2009) Protein thermodynamic structure. *IUBMB Life*, 61, 600–606.

Part II

System Logic and System Theory

Chapter 6

The Scientific Definition of the System

Summary: The system is a basic concept of science. However, until now, there has been scientific definition about it. In this chapter, we discuss its scientific definition on the basis of thermodynamics. According to this definition, a system represents one specific ensemble of matters with its specific structure: logic cycle.

In previous chapters, we have learnt the thermodynamics principles for the origin of natural order. However, this principle is rather theoretical; we cannot analyze the complex relationship between things.

In order to analyze the complex relationship between things, we must carefully study a most important question: the scientific definition of system concept on which the logic relation between different systems can be deduced theoretically.

We know that nature comprises many sorts of systems. Society is a complex system. The term "system" is utilized by all people, common people and scientists. All believe they understand the system concept and have no difficulty in using it.

However, the concept of system is very difficult to define. This may be beyond the ability of common people to comprehend. In this chapter, we study the general properties of the system and the complex discussion about the system will be found in the chapter of system logic.

1. System and Logic Cycle

In our discussion of the origin of natural order, the thermodynamics system formed in this process is a stable thermodynamics system.

In order to construct one stable system, a logic cycle is needed.

As we discussed in the previous chapter, advanced things are imposed upon restriction for the movement (or motion) of fundamental things.

Thus, advanced things are the outcome of the movement of fundamental things, and in another aspect, imposed restriction to the movement of fundamental things. It represents one type of a logic cycle.

This property can be seen clearly in protein science. The protein structure is the outcome of protein folding, and it restricts the movement of the components of protein.

In society, we also observe such a logic cycle. The leader is elected by society in diversified ways. The leader issues social rules for the common people.

In conventional views of mathematics, the logic cycle is not permitted.

Scientists do not distinguish the logic cycle phenomenon from the feedback regulation mechanism logically. In logic cycle phenomenon, the feedback is the conditional factor and it is not the parameter of

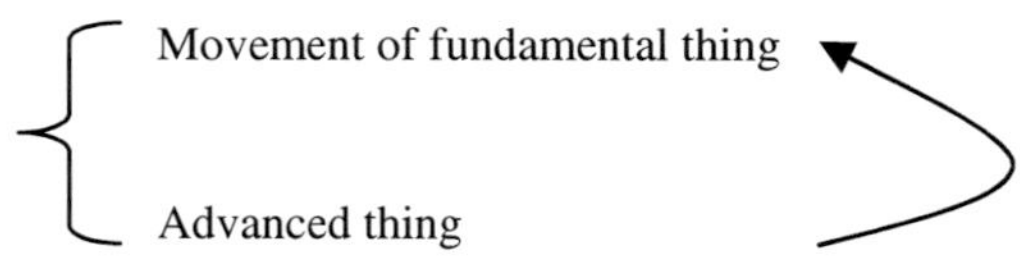

Figure 1. The logic cycle occurred in the origin of natural order.

The advanced thing is the outcome of the movement of fundamental thing, the advanced thing provide restriction to the movement of fundamental thing

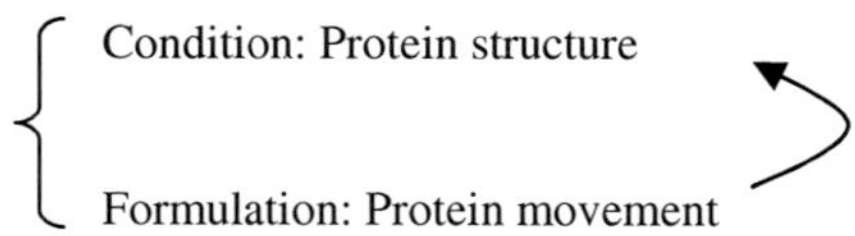

Figure 2. The relationship between protein structure and protein movement.

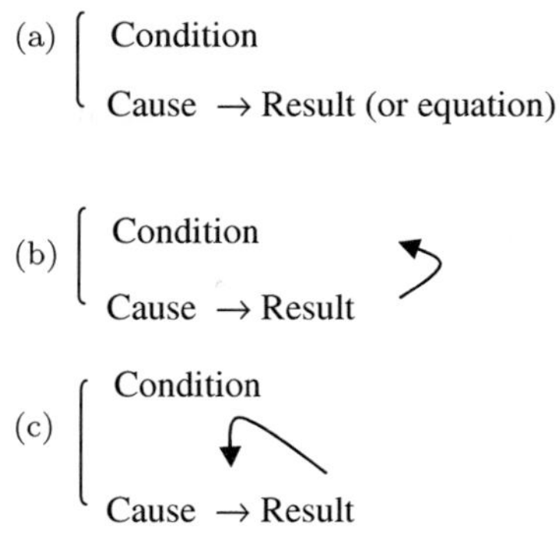

Figure 3. The logic cycle phenomenon and feedback regulation mechanism.

(A) Conventional views of mathematics. There is not a logic cycle.

(B) Logic cycle, the result is the factor (prerequisite condition) of the cause.

(C) Feedback regulation fashion. Feedback regulation mechanism does not alter the result/cause relationship. In contrast, the logic cycle has a great influence on the cause/result relationship.

an equation describing the relationship for cause/result. In contrast, in feedback regulation mechanism, the feedback property is the parameter of equation describing the relationship of cause/result. Figure 3C expresses logic cycle and Figure 3D expresses feedback regulation mechanism.

2. The Mathematical Expression of System Relationship

In order to gain a real understanding of a system, we must write it in mathematical language. By simulating the relationship between protein structure and protein movement, we formulate a system with two components in Figure 4:

Two highlighted features of these relationships can be clearly seen in Figure 4. The first is that the unified model written by formal logic cannot be developed. The second is stable states of a system, say ($Y>a$; $x>b$), may be generated in these relationships.

So, we get the mathematical definition of a system in Figure 5:

So, a more complex situation appears: the condition of a model is the output of this model. In traditional theory, the conditions are

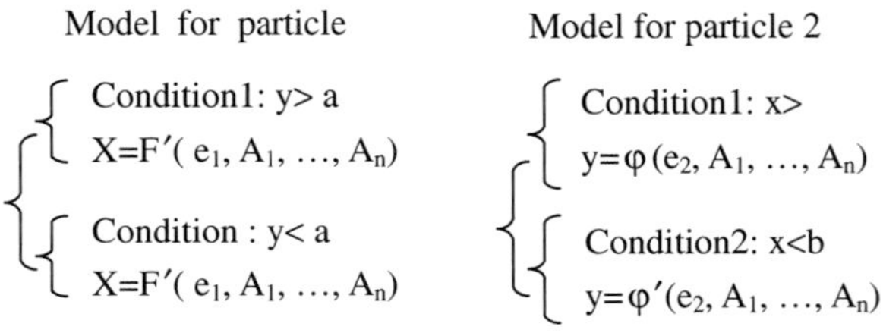

Figure 4. A system with two connected particles.

F and φ represent function. E represents element. A represents axiom. In this Figure, we see that a system comprised two connected particles (e1 and e2). The movement of a particle provides conditions for other particles. We can clearly see that these four axiomatic theories (or models) cannot be expressed by a unified axiomatic theory. The connections between two particles construct the structure of a system. For any property of the system, P=F(x, y, A). The P cannot be expressed by formal logic for there is complex relationship between X and Y.

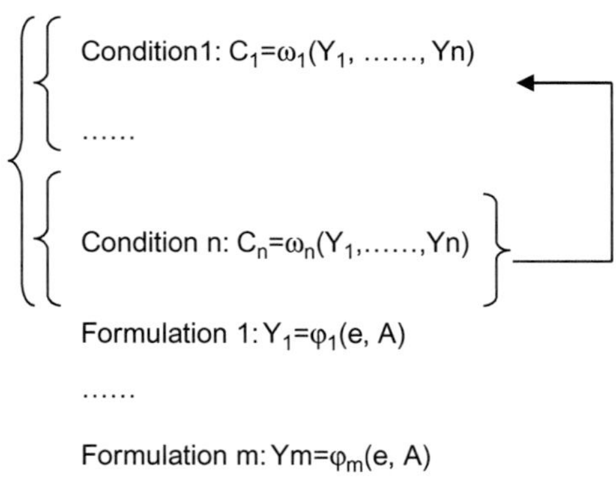

Figure 5. The mathematical definition of system.

In this system, the condition of a model is also the output of it.

the prerequisite for a paradigm and are independent of formulations of these models. This type of logic cycle has never been studied in traditional mathematics and physics. However, protein, a most complex thermodynamics system in physics, shows typical features of this system.

We formulate mathematical models for some systems in Figure 6.

(a)
IF Y≥ 0
Then Y=sinX
X=Y+π/2

(b)
IF Y> 0
Then Y=sinX
X=Y+π

↕

(c)
IF Y< 0
Then Y=cosX
X=Y+π

Figure 6. Some mathematical models of system.

A represses a stable system.
IF X $\neq$ 0, B and C repress two unstable systems that are oscillating.

These simple mathematical models have no physical meanings. We cite it here for the explanation of system.

3. Some Systems Fall out off the Scope of this Definition

Two types of thermodynamics systems fall out of the scope of this definition.

The first is a thermodynamics system of ideal gas that shows no particular structure and its behavior can be fully described by formal logic.

The second is a rigid body that shows a fully fixed structure and in some cases it can act as a particle, but the actual solid shows system behaviors in some cases. In fact, a study of these subjects produced conventional theories of physics, or mechanics and thermodynamics.

4. The Cooperativity Phenomenon

The change of a system shows cooperation phenomenon. The result can be easily obtained by studying Figure 4.

As the movements of components of a system are coupled, the system can only transform into another system suddenly.

The cooperativity of a system, in physics called phase transition (1), is a phenomenon of universal importance in physical, social, and biological systems, such as protein conformational change (2, 3), change of signaling transduction network (4). The metaphysical description of behaviors of many systems has been well established in history, but the real mathematical mechanism of it remains obscure. We need a fundamental mechanism for it to shape our thoughts about system biology, mathematics, physics, and complexity.

In biology, the cooperativity known as "mutualism" is one in which one species provides a haven or a metabolic advantage to another species. The logic feature of it is completely same as that shown in Figure 2.

A highlighted feature, which can be clearly seen in Figure 1 and this definition, is that the behaviors of a system cannot be described by elementary logic (or formal logic). This definition immediately offers a reasonable answer why we cannot define a system scientifically in earlier time.

Briefly, a system represents a complex system with its particular structure. There is logic cycle between its structure and its components.

As different systems structure differently, a system can transform into another system catastrophically. Thus, the system shows quantitative and qualitative change. The cooperativity phenomenon can only be seen in qualitative change. We have deduced the biophysical feature the protein structural (qualitative) and conformational (quantitative) change and it has been fully confirmed experimentally (5–7). Figure 7 illustrates the features of the system change.

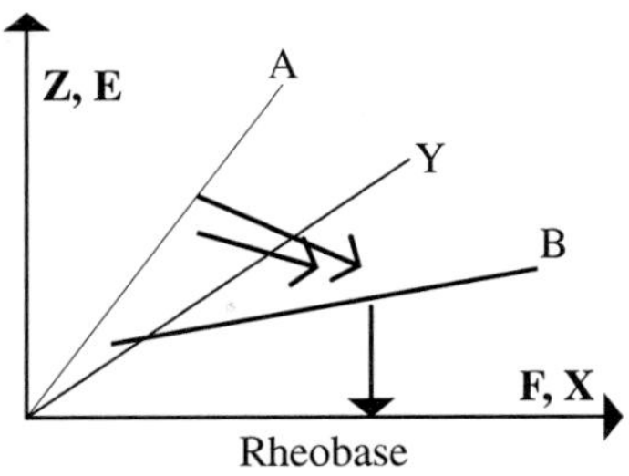

Figure 7. Qualitative and quantitative change of a protein.

X, Y, Z) Geometrical coordinate. E) The energy. F) Stimulation applied to a system. Rheobase) the minimum quantity of stimulation required for inducing structural change of a system. It represents the stability of a system. When a system acts on stimulation, it first contorts but remains its thermodynamic state (system A) then suddenly transforms into another state (system B). The change within system A is a quantitative change. The change between system A and B represents a qualitative change (structural change).

The system concept mentioned in this paper is completely compatible with the theory of origin of natural order or self-organization systems (8–12). A system can organize its particular structure itself and then can be self-organized. Protein folding provides a good example of it (3).

Recognition and study of systems should therefore both enrich and illuminate the phenomena of the physical state of substances, both biomaterials and chemical materials, in varied cases. The validation of our opinion should be further judged if this conclusion could be applied into other fields of physic (13–16).

References

1. Papon P., Leblond, J., Meijer, P. H. E. (2002) The Physics of Phase Transitions. Concepts and Applications. Springer, Berlin.
2. Perutz, M. F. (1990) Mechanisms of cooperativity and allosteric regulation in proteins. Cambridge University Press.
3. Horovitz, A. & Fersht, A.R. (1992) Co-operative interactions during protein folding. *J. Mol. Biol.* 224, 733–40.
4. Ferrell, J.E. (2002) Self-perpetuating state in signal transduction: positive feedback, double-negative feedback and bistability. *Current opinion in chemical Biology.* 6, 140–148.

5. Zhao, Q. Irreversible thermodynamic theory for protein folding and protein thermodynamic structure. *Progress in biochemistry and Biophysics.* **28**, 429–434 (2001)
6. Jiang, Y., Lee, A., Chen, J., Ruta, V., Cadene, M., Chait, B.T. & Mackinnon, R. (2003) X-ray structure of a voltage-dependent K^+ Channel. *Nature,* 423, 33–41.
7. Jiang, Y., Ruta, V., Chen, J., Lee, A., & Mackinnon, R. (2003) The principle of gating charge movement in a voltage-dependent K^+ channel. *Nature*, 423, 42–48.
8. Kondepudi, D. & Prigogine, I. (1998) Modern Thermodynamics: from heat engine to dissipative structure. John Wiley Press, New York.
9. Prigogine, I. & Stengers, I. (1997) The end of certainty: time, chaos, and the new law of nature. Free Press, New York.
10. Schieve W. C. & Allen, P. M. (1982) Self-organisation and dissipative structure: application in the physical and social science. University of Texas Press. Austin.
11. Ideker, T., Galitski, T., Hood, L. (2001) A new approach to decoding life: *Systems Biology. Annu. Rev. Genom. Hum. Genet.* 2, 343–372.
12. Sole R., Goodwin B. (2001) Signs of life: How complex pervades biology. New York: Basic Books.
13. Scandolo, S. (2003). Liquid-liquid phase transition in compressed hydrogen from first-principles simulations. *Proc. Natl. Acad. Sci.* USA. 100, 3051–3053.
14. Kasevich, M. A. Coherence with Atoms. *Science.* 298, 1363–1368 (2002).
15. Santen, L., Krauth, W. (2002) Absence of thermodynamic phase transition in a model glass former. *Nature.* 405, 550–551.
16. Anderson, V. J., Lekkerkerker, H. N. W. (2002) Insights into phase transition kinetics from colloid science. *Nature.* 416, 811–815.

Chapter 7

Logic Discontinuity and Continuity Between Different Levels of Hierarchical Structure

Summary: We show that the fundamental properties of matter cannot logically determine the properties of advanced matter. There is logical discontinuity between different levels of hierarchical structure, which reveals the theoretical limit of scientific model or paradigms. It also shows that we cannot analyze advanced matter by fundamental properties alone.

Nature, as well as biology, is a developing system with a hierarchical structure, which cannot be described by one formal paradigm. Any axiomatic theory can only describe one detailed scientific subject, such as an element (particle) of mechanics. One substance may take different roles within different models, which describe detailed properties of substance at given levels, and logic discontinuity may be observed. The study of this phenomenon is of great significance for understanding of the complex connections of the nature. Mathematical analysis shows that the bioinformation of some genes can act as one element of one axiomatic theory and can only act as a conditional factor in another axiomatic theory.

We have leant that nature, as well as many of its components, show a hierarchical structure, which is generated in the irreversible thermodynamic process of natural change. The general mechanism for it has been carefully discussed.

Another question is the relationship between things at diversified levels and the logic relationship between different levels. In the following parts of this book, we show the significance of this.

Some scientists have studied these questions with no clear-cut resolution. Several approaches to this subject are complexity, chaos and the logic limit of an axiomatic theory (1–9). However, all such approaches are targeted to the relationship between concrete matters. The relationship between axiomatic theories remains unexplored.

In order to formulate a clear logical relationship, we must study the axiomatic theories mathematically. Normally, an axiomatic theory at one level could be theoretically developed and it would describe the properties of substance on a given level. Therefore, we must study the relationship between these different axiomatic theories.

1. The Mathematical Analysis of Axiomatic Theories

We can logically formulate the relationship of different axiomatic theories at different levels of developing system as follows:

Six highlighted features can be clearly seen in this Figure: a) the effect of some elements of the first axiomatic theory can only be a conditional factor within the second axiomatic theory. Its property cannot logically be expressed in the second axiomatic theory. The logic discontinuity occurred in the transformation from axiomatic theory 1 to axiomatic theory 2. b) These two theories cannot be annexed into one unified axiomatic theory. c) The relationship between two theories (axiomatic theory 2 and 3) cannot be deduced logically from anyone of two theories and their combination. d) Different value of $Y_{1.1}$ can select different models as following the axiomatic theory and the effect of it can be shown in this particular process. e) The (a, b) can be recognized as the stability (or definition domain) of the axiomatic theory 2. f) There is incompatibility between axiomatic theory 2 and 3. Either of them can act as following axiomatic theory for axiomatic theory 1, but they cannot coexist for a given case.

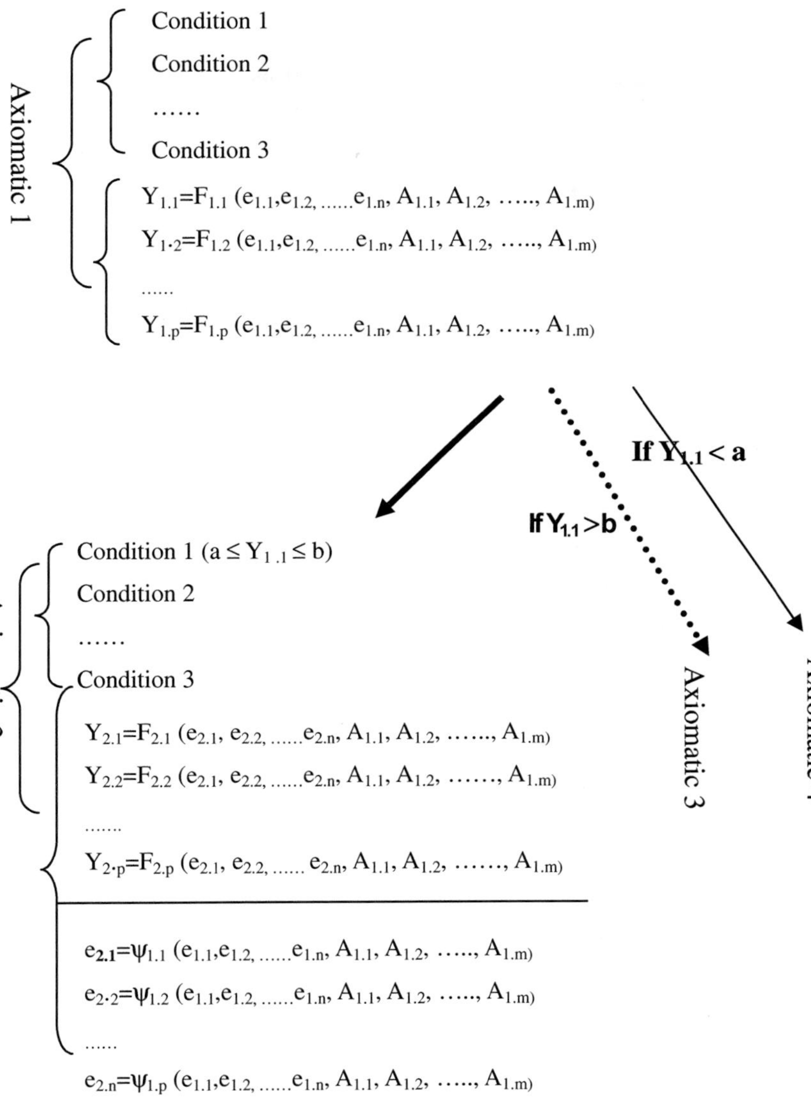

Figure 1. The relationship between different axiomatic theories.

Where F and ψ represent Function, e represents elements of axiomatic theory. A represents axiom of the axiomatic theory.

Several conclusions could be deduced logically: 1) the unified axiomatic theory describing all features of matter which shows hierarchical structure cannot be developed. 2) The unified theory of physics, which covers all areas, cannot be developed theoretically. 3) The unified theory of genetics (biology) cannot be developed (10–12).

These properties agree with features of bioinformation flow from gene to biological function well. We have known that fact: Our ability to think is the product of neural signal conduction activities. The genes of the ion channel influence our ability to think, but do not logically determine our thought. The individual with gene mutants of the ion channel has a tendency toward neuropathy (13). The environment can also be a conditional factor of axiomatic theory 2; so biological function can be influenced by environment.

2. The Diagrammatic Notes for Above Analysis

Some people may find it difficult to understand mathematical language, so its relationship is demonstrated in Figure 2.

As different axiomatic theories are validated in different conditions, they cannot be unified into one general axiomatic theory.

The relationship between 2 and 3 also construct another model. However, not all possible scenarios are listed here.

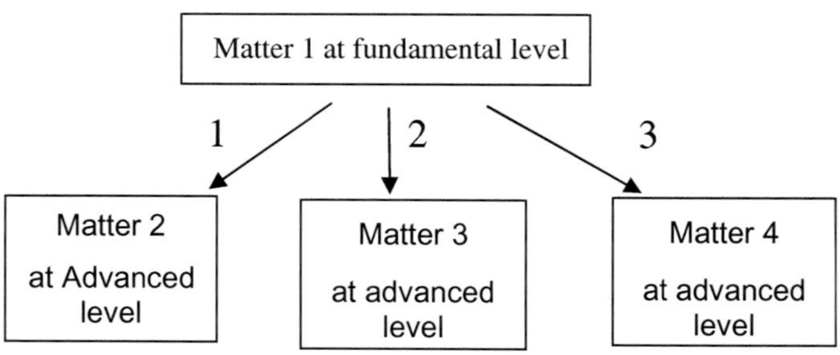

Figure 2. The relationship of matters.

The outcome of matter 1 at a fundamental level generates the condition for advanced matter. Condition 1 results in the origin of advanced matter 1. The condition 2 results in the origin of advanced matter 2. Condition 3 results in the appearance of advanced matter 3.

3. The Logic Continuity and Discontinuity Between Hierarchical Structure

Based upon the above discussion, the following principles for the relationship between matter has been established.

Principle 1: For any one property of fundamental matter, many models (axiomatic theories) could be developed and within these models the properties fundamental matter cannot be merged. This represents the logic discontinuity between materials of hierarchical structure.

In Figures 1 and 2, we see that the property of matter 2 and 3 (formulation or logic of axiomatic 2 and 3) has nothing to do with any properties of matter 1. In our discussion of the property of matter 2 at advanced level, it is unnecessary to refer the properties of matter 1.

Principle 2: For any properly of fundamental matter, many models could be established and within these models the properties of fundamental matter could be logically established. We can study their effect within this model.

The properties of matter 1 take effect when we study the relationship between matters 2 and 3 (axiomatic 2 and 3). It is helpful here to consider that the relationship between matter 2 and 3 constructs another axiomatic theory.

This way, the effect of matter 1 could emerge into an advanced level, is called logic continuity among different levels of hierarchical structure of maters. We can judge the existence of fundamental matter.

Principle 3: The hierarchical structure cannot be explained by one axiomatic theory written by formal logic. This principle is of great significance in our understanding of nature. Due to it, the unified theory of physics cannot be developed; the unified axiomatic theory of biology, which can explain all things, cannot be developed. So, we consider our book as biology: a contextualized axiomatic theory.

A contextualized axiomatic theory, cannot explain all properties (processes) of biology. But it can provide a theoretical framework for the structure of biology, and this framework is logically compatible with all knowledge of biology.

With this principle, our knowledge of science, in its most fundamental essence, is constructed. By introducing the conditions provided by experimental science, the unified structure of our knowledge could be obtained.

4. The Unity of Things and the Specificity of Thing

The principles mentioned above immediately offer a theoretical explanation to the unity of nature (things of it) and specificity of matter.

If we study the relationship between matter at diversified levels, we must ignore some of the specificity. If we study the specificity of matter, the unity of nature will not be addressed in our scientific models.

It provides a reasonable explanation for why experimental scientists have no interest in the unity of nature for they study the specificity of things, and the unity of nature has nothing to do with their scientific subject.

A conclusion may be obtained from these principles. The synthesis of basic principles of matter at fundamental levels cannot result in the real understanding of the science for the relationship between things varying under different model. We do not discuss it here.

5. Examples

The concept of logical discontinuity provides explanation to many questions. But we will not discuss these questions here. We cite several examples for better understanding.

We all know that water supports our physiological activity. Disease will occur if the water regulation mechanism does not work well. The lack of water can impair our spirit, and even more some biological functions. But our thought is not logically related to properties of water. Being human cannot be attributed to any properties of atom, protein, and gene.

So, although the life is made up from physical materials, it has particular properties which are not explainable by fundamental properties of materials.

References

1. Kondepudi, D. & Prigogine, I. Modern Thermodynamics: from heat engine to dissipative structure. (John Wiley Press, New York. 1998).
2. Prigogine, I. & Stengers, I. The end of certainty: time, chaos, and the new law of nature. (Free Press, New York 1997).
3. Kauffman S. A. The origin of order: self-organization and selection in evolution. Oxford University Press, New York, 1993.
4. Zhao, Q. Irreversible thermodynamics theory for protein folding and protein thermodynamics. *Progress in Biochemistry and Biophysics* 28(3) 429–435 (2001).
5. Darwin, C. On the origin of the species by means of natural selection or the preservation of favoured races in the struggle for the life. (John Murray, London, 1859).
6. Kline, M. Methematics: The loss of certainty. (Oxford University Press 1980 London, 1980).
7. Barrow, J. D. Impossibility: the limits of science and the science of limit. (Oxford university Press London, 1998).
8. Lorenz, E. N. The essence of Chaos. (University of Washington Press, Washington. 1993).
9. Ricard, J. Biological complexity and the dynamics of the life processes. (Amsterdam Elsevier 1999).
10. Morgen, T. H. The theory of the gene. (Yale University Press. New Heven, 1926).
11. Fairbanks, D. J. & Anderson, W.R. genetics. (Thomson Publishing Company, 1999).
12. Crick, F. H. C. Central dogma of molecular biology. *Nature* 277, 561–563 (1970).
13. Hunter, A. J., Nolan P. M., Brown S. D. M. Towards new models of disease and physiology in the neurosciences: the role of induced and naturally occurring mutations Hum. Mol. Genet. 9: 893–900 (2000).

Chapter 8

System Logic

Summary: All things in nature are systems, but the principles of system have not been formulated mathematically. Here, we formulate a standard paradigm for systems. Basic concept: 1) a system is complex system with its particular structure and it cannot be described by element logic (or axiomatic theory, the traditional logic of mathematics). 2) The relationship of systems can be only described by system logic. The relationship between system logic and element logic are discussed. It provides a new way for our understanding of complexity of things.

Until now, our mathematical logic about matter and nature has been limited to element logic (Axiomatic theory) to be fully expressed within formal logic, which was first proposed by Aristotle (1–3). In this method of scientific research, the elementary substance of the axiomatic (or model) was first isolated (or separated from nature), after the properties of this elementary substance were studied. The general properties of this elementary substance, for example the particles as theorized by Newton, were then renamed as basic scientific principles, which could then be utilized to explain nature. Thus, many axiomatic theories of axiomatic theories could be developed, but the relationship between them remains obscure. Modern physics is purely established based upon this scientific research method, and to date, great strides have been made in this field. But the ability of this sort of scientific logic has not been developed sufficiently to explain biological and social questions in isolation.

Gödel, a German mathematican, first confirmed the theoretical flaw of axiomatic theory. This creates doubt with regard to traditional logics ability to explain nature and itself (4, 5).

The theoretical flaw of this research method is simple and clear if we study the thinking method. In Euclidean geometry, a point has no part or infrastructure, and have not any length, area, volume, or any other dimensional attribute, but a line segment is composed of points. There is contradiction within this definition. Similarly, in Newton theory, a particle has no volume and infrastructure, but it has weight and other properties. In views of natural science, it is impossible. Thus, Newton's theory is not the truth, but a scientific model which is validated under some cases.

This conceptual evolution of science is strengthened by several approaches, which include the complex system (complexity) (6–11), irreversible thermodynamic process (12–14), in fields of mathematics and natural science, see, biology and social science (15–16). They share similar philosophical opinions but differ in that they select different models to demonstrate it and the consensus of these approaches remains to be confirmed in future.

The word "system" is a common term in everyday language and has attracted much attention both among lay people and scientists working in diverse scientific fields (8–10). Some remarkable achievements have been made in understanding the nature of systems. (4–10, 17). However, the scientific logic surrounding systems and scientific definition of it has not been fully explored or developed. Our comprehension of systems is not established based upon a theory of system, but mostly reflects our doubt about the traditional knowledge of mathematics.

Our interest in the study of these subjects arose from our research on protein folding and protein thermodynamic structure (10) and the scientific explanation of Traditional Chinese Medicine (17). In this research, we found that the properties of proteins cannot be logically reduced into some features of amino acid residues, the basic units of protein. We found that a protein is not global thermodynamic system, but is composed of many sub-systems, which are called potherse. This enforces us on studying the mathematical logic of system (system logic) theoretically.

Firstly, we found it is impossible to isolate a fundamental (element) property of a protein and all properties of proteins can be deduced based upon it. Finally, we found that system logic differs from element logic in principle. The system cannot be described by element logic. In this paper, we discuss the mathematical principles of system.

1. Element Logic and Formal Logic

In the western world, the first thinker to gain real understanding of scientific logic was Aristotle, whose work provided the foundation of modern scientific logic (1–3). He developed the concept of axiomatic theory. Euclidian geometry was the first axiomatic theory in history, which made possible the extension of Aristotle's hypothesis into particular fields of science. Much later, Newton developed his theory of Mechanistic, which was the first axiomatic theory to deal with the world of natural science. Mechanistic has had a profound impact upon the history of science. As different as they are, traditional equilibrium thermodynamics, quantum theory and relativity theory are all based upon this brand of scientific logic, but with some variations.

One common feature of these theories is that they all involve absolute concepts. Theoretically speaking, all questions should be able to be explained by employing these absolute concepts, although in practice, this turns out to be impossible. Because the scientific concepts within these systems of element logic are absolute, element logic can be expressed in terms of symbolic logic.

Over time, scientific researchers have made enormous progress in expanding our understanding of nature and science. But with this understanding comes a new challenge: How do we explain, in logical and scientific terms, the underlying relationship between the myriad scientific theories in the fields of natural and social science? How do we unify these theories to gain some understanding of the bigger picture, the fundamental features of nature and science? Thus far, our understanding of nature has been fragmented, simply by virtue of the vast number of different theories in existence today. In addition, of course, there are still many questions that have not been satisfactorily resolved within these theories, and many questions remain that seem incomprehensible within our present worldview.

It is well known that element logic is not particularly well equipped to handle certain social and biological questions. Although the traditional scientific method has been widely applied in biology and the social sciences, it remains limited as a tool for scientific research in these fields. Traditional scientific principles and logic alone cannot explain social and biological questions adequately enough.

One might pose the question of whether the logic to thoroughly explain objects and natural phenomena can ever truly be abstracted from our observations of nature and natural events. And if this is not possible, what sort of logic do we use to explain them? There are many questions concerning this question. For example, can we find the fundamentals of human nature? We have developed no satisfactory logical explanation in past. Scientists have struggled with these questions in all fields of science and culture, and disagreement has raged almost since the beginning of human history.

2. The Definition of the System

We have studied the scientific definition of system in the previous chapter. We study the general principle of system here.

It could be hypothesized that any one component of a system can be fully described by formal logic. The movement of one particle provides a conditional factor that restrains the movement of other particle.

With these connections, we could not annex the models that describe the single particle into one unified model. The reason is rather simple; we cannot handle conditional factor logically. All mathematicians know it.

In order to describe this system, more axiomatic theory, which is meaningful in a limited area, is required.

Of course, the actual case of a protein is more complex than what we have studied here. But a definition of system could be deduced from this study.

A system is comprised of many components with its particular structure. The behavior of a system cannot be deduced from its components as properties of its particular structure cannot be defined by the properties of its components. The properties of a system cannot be deduced from properties of its components.

In protein science, the movement of protein components is controlled by protein structure. The protein structure cannot be deduced logically from properties of protein components, see amino acid residues.

Thus, a system comprises two different parts: structure of a system and components of a system, which cannot be expressed by one unified axiomatic theory. We cannot describe all features of a system by an axiomatic theory.

3. The Properties of System

Combining the above analysis and thermodynamic structure theory of a system (10), the properties of system were given as following:

The structural change of a system and qualitative change: The components of a system can be structured differently. As two systems are structurally different, the transformation between them can be only processed catastrophically and it produces the qualitative change of a system. We have seen many examples of it. In physics, the electron from one orbit to another orbit happens suddenly. We have predicted protein structural change of the ion channel and its biophysical features (10) and it has been confirmed experimentally (18–19), see Figure 4 of Chapter 4.

Quantitative change of a system and physical variables of a system: A system can tolerate some degree of stimulus, and permit some degree of change of system properties, without inducing structural change to a system. If the quantity of a variable is increased enough that it destroys the stability of the system, the fundamental nature of the system will change. It is possible for one variable to vary freely to some degree. But because the variables of the system are not fully independent of each other, they cannot vary freely beyond certain limits.

A system has unlimited variable. When one system interacts with another, two systems must be deformed in a degree to match up its optimum interaction. Thus, a third system is produced. This system is compatible with other two systems, but its properties cannot be deduced from the properties of parent systems for it can structure itself. **This is a highlighted feature of a system**. The conclusion is that there is no axiomatic theory that can describe all properties of a system.

Thus, a system can behave differently when it interacts with different systems. The behavior of a system becomes one physical variable in specific cases. So, a system has unlimited variables (or properties). It must stress here that the property of a system can be only expressed when it interacts with other systems. There is no absolute property of a system. This makes a key difference between an elementary particle and a system.

In protein science, a protein can interact with different proteins in different ways.

In some specific cases, a system can act as a particle. When we disregard the inner movement of a system, or theoretically eliminate the effect of the inner physical movements of a system by using certain scientific methods, a system may be seen to act as a single particle. In this case, certain properties of the system can be analyzed by using element logic. Therefore, element logic can sometimes approximate system logic to describe certain natural objects and phenomena.

If we wish to analyze an object or phenomenon in terms of mechanistics, we can apply the concept of a "rigid body". Once a substance has been recognized as a rigid body, its behavior can be described by mechanistics.

In the field of protein research, scientists studying the osmotic pressure of certain solutions have found that protein acts as one particle within a solution (20).

The fact that a system can behave as a particle in certain specific cases enables us to analyze some properties of a system by using element logic. And therein lies the explanation for why science (particularly physics and mathematics) was able to develop in history based entirely upon element logic.

4. The System Logic

The above analysis finally made us formulate a system logic, which could describe relationships between systems.

Principle 1: The relationships of two systems construct a new system. The new system possesses entirely new features that have not been expressed by any other systems.

In the field of geo-politics, we can see one example of a new system arising: The United Nations, which was established in order to resolve disputes and regulate relations between different countries.

In biology, the regulation of the biological organs is undertaken, in part, by the activities of the central nervous system.

In most cases, such new systems are rarely as concrete as the examples mentioned above. They are more likely to take the form of "dummy systems", such as systems that exist between dissimilar languages - English and Chinese, for example.

Principle 2: Between two systems, many models could be established, and in these models the relationship between components (properties) of systems can be written by formal logic.

However these models do not represent the entire system relationship as a whole, nor of any combinations between them.

The reason is rather simple. The different models are obtained based upon different conditions, so we cannot unite these different conditions in one scientific theory. In addition, some relationships between systems will not be covered in these relationships.

For example, while we can find corresponding relationships between certain English words and their Chinese equivalents, we cannot understand English based upon these relationships alone. There are many characteristics of the English language, such as idiomatic expressions, that cannot be explained based on this relationship between English and Chinese.

In scientific research, varying scientific theories arise in the course of different scientific research. There is no fixed relationship between this type of theory and the actual facts we may observe. But when we write a book, naturally, we must seek out the relationships between them.

Principle 3: Among these models written in formal logic, which describe the relationship between systems mentioned in principle 2, some models are incompatible with each other.

The logic reason is rather simple. If all models are compatible with each other, a unified axiomatic theory could be developed, at least theoretically but is not permitted logically.

For example, because language itself is changeable, so is the relationship between languages (say, English and Chinese). Although English will never change so radically that it actually *becomes* Chinese because they are not compatible with each other logically, we can observe that over time, English, particularly American English, is absorbing more and more Chinese idiomatic expressions.

In most case, we are forced to deal with complex systems (relationship between systems) without knowing the detailed structure of a system. As unified axiomatic theory cannot be developed, some conditional models are valuable although these models are not validated in all cases.

By applying principles noted above, we could get the following conclusions about properties of systems.

- Any property of a system (or one relationship between systems) is attributed to a sub-system of it. The formal logic could be developed to describe this particular property under specific conditions, but the validity of it is limited.

- Different properties are attributed to different sub-systems.
- No unified theory, which could explain all properties of a system, can be developed.

In conclusion, there are many systems and their properties cannot be described by formal logic. This concept has been successfully applied into field of protein science. It may find its way into other fields of science, such as physics, mathematics, as well as social science in future.

5. The Fundamental Unit of Thing Does not Equal to the Elementary Concept of Axiomatic Theory

For complex systems, such as society and protein, the individual and amino acid residues are basic units. Most people believe that the behaviors of society and protein may be explained by one axiomatic theory and these basic units act as elements of this axiomatic theory.

Although such axiomatic theory has not been developed, some people think that it will be developed when more data and theoretical study have been made.

Among the consideration of complex relationship between things, such theory cannot be developed. The mathematical demonstration will be offered in following chapters.

6. Which Questions Could be Resolved by System Logic

Many questions of science and biological question could be answered by system logic. We cite example of it here, but we will not step into detailed case for many questions of it for it falls out the scope of this book.

1. The relationship between different scientific theories (or models).
2. The relationships between complex things.
3. The relationships of things between different levels of a hierarchical structure.

A scientific theory or model of experimental science represents a specific system. So their relationships could be analyzed by system logic.

In biology, a protein particle, DNA particle, a biosignal network and a cell (and biological organs) represent different systems. Their physiological relationship could be analyzed by system logic.

In physics, all things (atoms, elementary particles, protons, electrons) represent different systems. But we do not study these physical questions here.

In fields of philosophy, dialectics could be accepted by system logic.

References

1. Katz, V. J. (1998) A history of Mathematics. 2nd Ed., Addison-Wesley.
2. Paul, T. (1999) Logic. Routledge, New York.
3. Chew, G. F. (1974) Impasse for the elementary particle concept. The Great Ideas Today. Chicago.
4. Kline, M. (1980) *Methematics*: The loss of certainty. Oxford University Press 1980 London.
5. Barrow, J. D. (1998) *Impossibility*: the limits of science and the science of limit. Oxford University Press London.
6. Lorenz, E. N. (1993) *The essence of Chaos*. University of Washington Press, Washington.
7. Ricard, J. (1999) *Biological complexity and the dynamics of the life processes*. Amsterdam Elsevier.
8. Sole, R. and Goodwin, B. (2001) *Signs of life*: How complex pervades biology New York: Basic Books.
9. Ideker, T., Galitski, T. and Hood, L. (2001). A new approach to decoding and life. *Systems Biology. Annu. Rev. Genom. Hum. Genet.* 2, 343–372
10. Zhao, Q. (1996) The physical movement of macromolecules and the physical process of signal transduction. *Chinese J of Rehabilitation Theory & Practice* 2, 135–140.
11. Zhao, Q. (2001) Irreversible thermodynamics theory of protein folding and protein thermodynamics (English paper). *Progress in Biochemistry and Biophysics* 28(3), 429–435.
12. Kondepudi, D. and Prigogine, I. (1998) *Modern Thermodynamics*: from heat engine to dissipative structure. John Wiley Press, New York.

13. Prigogine, I. and Stengers, I. (1997) *The end of certainty*: time, chaos, and the new law of nature. Free Press, New York.
14. Willim C. Schieve and Peter M. Allen (1982) *Self-organisation and dissipative structure*: application in the physical and social science. University of Texas Press Austin.
15. Boher, N. (1934) Atomic physics and the description of the nature. Cambridge University Press: Cambridge.
16. Wiene, N. (1949) *Cybernetics John Wiley & Sons Inc.* New York.
17. Zhao, Q. and Wang, A. (1999) The scientific interpretation of traditional Chinese medical theory. The proceeding of excellent medical papers of Chinese's (military publisher, Beijing) pp. 3235–3246.
18. Jiang, Y., Lee, A., Chen, J., Ruta, V., Cadene, M., Chait, B. T. and Mackinnon, R. (2003) X-ray structure of a voltage-dependent K^+ Channel. Nature 423, 33–41.
19. Jiang, Y., Ruta, V., Chen, J., Lee, A. and Mackinnon, R. (2003) The principle of gating charge movement in a voltage-dependent K^+ channel. *Nature* 423, 42–48.
20. Lenhinger, A. L. (1984) Principles of biochemistry Third ed., Worth publishers Inc. New York.

Chapter 9

Isolating Logic Relation Between Complex Systems

Summary: When complex is merged, it is impossible to find absolute model which can account for all matters. In these cases, it is valuable to find conditional models and the logical relationship between different matter could be constructed within these models. These isolated models have great impact on our understanding of nature.

Although the fundamental mechanism of the origin of natural order is simple and the complexity concept is unnecessary, the hierarchical structure of matter generates complexity. We have learned that the unified axiomatic theories that can cover all areas and questions of hierarchical structure cannot be developed, but it is nonsense to claim complexity alone without any method to resolve the complexity. We must seek the conditional model within these models where the elementary logic is validated.

We must also isolate scientific logic among the complexity or complex relationship between things.

1. Isolation of the Linear (or Direct) Logic Between Systems

When we are faced with the full complexity (or uncertainty) of science, and we find ourselves at an impasse where we lack explanatory power, an alternative scientific method for dealing with practical problems should be developed. In the world of science, the isolation of linear relationships between systems is one of the most powerful such methods at our disposal.

The interactions between systems can be expressed only under specific conditions. The relationship between systems can be described by using conditional logic. This means that when we study the interaction of systems, different conditions will give rise to different systems of logic. We refer to this conditional logic as the "isolated logic of systems". It is a logic that can be isolated under specific conditions and structured within a particular scientific paradigm. The interaction of systems manifests itself differently under different conditions. By combining system logic and the theory of the origin of natural order, we obtain the following conclusions:

1. There is always one scientific paradigm, at least, in which a linear (or direct) scientific logic can be formulated to describe the interactions of different properties of related systems. It is only in these particular models that scientific logic can be expressed within forms of symbolic logic.
2. There are many scientific paradigms that can be utilized to describe the interaction of the different properties of a system. In these paradigms, the interaction of the system can be described using a system of linear logic.
3. There are some scientific paradigms that are not compatible with each other. The reason for this is simple: when different scientific paradigms are isolated under different conditions, they cannot concur. For example, it is impossible for "condition A" and "condition not-A" to logically concur.

In system logic, if a system cannot be recognized in terms of elementary logic, all scientific logic becomes conditional logic. In this case, scientific logic can be isolated by a variety of different scientific research methods. Because these systems of logic are valid only under certain conditions and are meaningful only within a specific paradigm, they cannot be utilized to represent the entire relationship of systems. This is also true for any combination of these systems of conditional logic.

Conditional logic of this sort can be expressed in terms of formal logic. System logic, however, cannot be expressed in terms of formal logic.

One well-known advocate of chaos theory, Lorenz, said once during a lecture that a single butterfly fluttering its wings might bring about a tornado in Texas. This quote, which highlights the extreme sensibility of chaos theory to disturbance, has become something of a catchphrase in chaos theory.

Analyzing this quote in terms of system logic reveals that it represents a certain isolated (or conditional) logic. But carefully study reveals that there are some scientific questions that cannot be resolved within this idea. Theoretically, one butterfly beating its wings could induce a tornado, but only under very specific conditions. If these conditions manifested themselves in reality, it follows that other small disturbances could also induce a tornado. Therefore, it is these conditions themselves, rather than the beating of a butterfly's wings, that act as the logical determinants of a tornado. In that case, would it not be more useful to study the specific conditions that make this possible, rather than speculating about the effect of the butterfly beating its wings?

The isolation of conditional logic (or relationships) has been widely utilized by scientists, whether they are cognizant of it or not.

In physics, the abstraction of scientific concepts in different scientific theories has been proven to be a useful method in scientific research. The very divisions within the world of science, divisions into separate and diverse scientific fields, are proof positive of the fruits of our scientific research. As most people are aware, most physical principles are meaningful only within a limited area of physics. Even Newton's First Law is expressed conditionally.

In biology, the isolation of the linear relationships between genes and biological traits is the foundation underlying modern genetics. In my opinion, the great achievements of Mendel were obtained by precisely this method.

In society, the method of isolating linear relationships to explain events and phenomena is so widespread that it has become common sense. Ask any politician and he or she will certainly recognize the fact that all events happen under certain specific conditions.

2. Isolating Conditional Logic in Areas of Thermodynamics

In university, the student learnt that the principles of thermodynamics come from our study on ideal gas. It can only describe properties of ideal gas. Theoretically they can only approximately describe the behaviors of actual gas, such as oxygen.

Gradually, the student found that the principle of thermodynamics can describe the behaviors of chemical reaction occurring in solution, and more, they finally found that principles of thermodynamics can describe behaviors of solid material.

The student cannot find the answer for why the principle of thermodynamics cannot describe such complex materials.

In protein science, the student was told that a protein acts as one particle in solution. The osmotic pressure of solution containing protein can be calculated by particle theory. In another aspect, the textbook of biochemistry also mentioned the inner movement of protein. Why does a protein, a complex material, act as particle? In which case can we consider it as particle? And in which case can we not consider it as particle? Are there rules to rely on?

The teachers also have no answers for these questions? They can only tell students that these principles are validated in these cases for more scientific prediction based upon experimental testing.

The answer comes from system logic. For any concrete measuring (or property) of the thermodynamics system, a thermodynamics subsystem can be isolated from a complex thermodynamics system and the behaviour of this sub-system can be described by principles of thermodynamics.

One question raised by this division is the relationship between these sub-systems or the relationship between the parameter of these sub-systems.

This question is very complex. We will discuss it in protein structure theory and will provide an example.

We know that the living body is at a nonequilibrium thermodynamics state and that the concept of temperature is nonsense in fields of nonequilibrium state. But we measure the body temperature daily. What is the matter?

One answer, most doctors said, is the temperature measured is an approximation. This may be correct in one aspect. The right answer may be that some processes of living body are at equilibrium state and some processes are at a nonequilibrium state and the temperature may reflect thermodynamics properties of equilibrium states and processes. It is an isolated concept, not the absolute concept.

3. The Principle of Free Coordinates of Thermodynamics

In view of thermodynamics, a thermal system represents a coordinate (freedom, or free dimension) of thermodynamics. Please make in mind that the concept of a thermal system refers to thermal systems obtained from our studying of thermodynamic structure of material, and it may make confusion if we apply it into other cases without careful consideration.

It states that for any dimension of thermodynamics systems, natural or artificial, the partition function of a thermal system is same.

Therefore, we can describe any properties of thermodynamic system as that of any thermal motion as we have learned in textbook of physics.

It makes us handle complex question in a simpler model.

We will discuss this principle in protein thermodynamics structure in following chapters.

Part III

Axiomatic Theory of Biology

Chapter 10

Irreversible Thermodynamic Structure Theory for Protein Folding

Summary: When polypeptide is synthesized *in vivo*, the nascent peptide begins to fold itself and finally a folded protein with full biological function is produced. Theoretically speaking, the protein folding is irreversible thermodynamic process and we discuss its principles here. Key thoughts: 1) nature of protein folding can be described by laws of irreversible thermodynamics; 2) a folded protein may be at metastable state in view of thermodynamics; 3) the product of protein folding is protein thermodynamic structure; 4) the Gibbs free energy is not the controlling factor for protein folding.

If not all, at least most biological functions are carried out by proteins *in vivo*. Thus exposition of relationship between protein structure and biological function is crucial for our understanding of biology. The polypeptide, which is synthesized in ribosome, has no biological function unless it has folded into a tightly packed or folded protein with its unique three dimensional structure. Physically speaking, the protein folding is a typical process for the origin of natural order (1–3). In view of protein conformation, the polypeptide represents nothing and folded protein represents everything. In this chapter, we will discuss basic thermodynamics principle of it.

The Anfinsen hypothesis for protein folding had been widely known over the world (4, 5). This model was established on the experimental tests that proteins can be unfolded and refolded via equilibrium thermodynamic way (or reversal process of thermodynamics) in solution containing urea or guanidine hydrochloride (GmHCl) (4, 6). According to this theory, a folded protein is thermodynamically most stable state under physiological condition.

However, many scientists are skeptic to Anfinsen's theory (7–11). We have summarized following evidences and reasons which is not compatible with Anfinsen's hypothesis: 1) although unfolding and refolding of a protein can be observed experimentally in solution containing urea or GmHCl, however, they are not quasi-static process and the conclusion of Anfinsen cannot be strictly deduced from results of such tests. 2) The equilibrium process of thermal denaturation and renaturation of a protein has never been observed and there is no physical reason to assert that native state of a folded protein is more stable than aggregation state in view of thermodynamics; 3) the prion biology has provided a powerful evidence that native protein can fold into a kinetics stable structure, rather than thermodynamics stable structure (7); 4) a widely accepted concept is that protein folding is kinetics-control, not thermodynamics-control process, this recognition fundamentally differs from equilibrium thermodynamics hypothesis for protein folding (1, 10, 11); 5) *in vivo*, a polypeptide can fold into active state of a protein within a short time, say 2–5 minutes, however, the timescale for equilibrium thermodynamics folding and unfolding is larger, say 0.5–12 hours (4, 6); 6) *in vivo*, the environment for protein folding remains unchangeable and there is no chemical or physical gradient between final state and initial state for protein folding, which is the prerequisite for equilibrium thermodynamics hypothesis for protein folding; 7) Levinthal has predicted that a nascent polypeptide cannot find its most stable state of thermodynamics in a limited period of time by way of random collision mechanism (8); 8) some protein can fold with assistant of chaperone (12). Among the consideration of above reasons, most scientist now believe that protein folding is irreversible process of thermodynamics although they use different terms, such as kinetics-control process (1, 9, 10), non-equilibrium process (11), to describe it. According to current evidences, the folding of small or globule proteins is self-assembly process and follow a well-ordered pathway (13).

1. The Time Scale and Thermodynamics Criterion of Irreversible Processes

When we analyze multiple irreversible processes of a thermal system as those represented in Figure 1, the first observation is the different time scales of different thermal processes. The thermal systems will show different configurations at different points of time and finally it reaches to an equilibrium state of thermodynamics. At equilibrium state of a thermal system, the probability of every state can be precisely calculated by partition function of thermodynamics. In contrast, at any time of irreversible process, the probability of thermal state cannot be calculated by partition function of thermodynamics. Thus, for a system subjected to irreversible change, the time is an important parameter which characterizes the nature of a system. This is of particular importance in the study of the relationship of many irreversible processes, such as that occurs in protein folding.

We can also analyze the effect of time in Figure 2. When a system is subjected to change irreversibly, many–subsystem, which are always at equilibrium state, appears within this larger system.

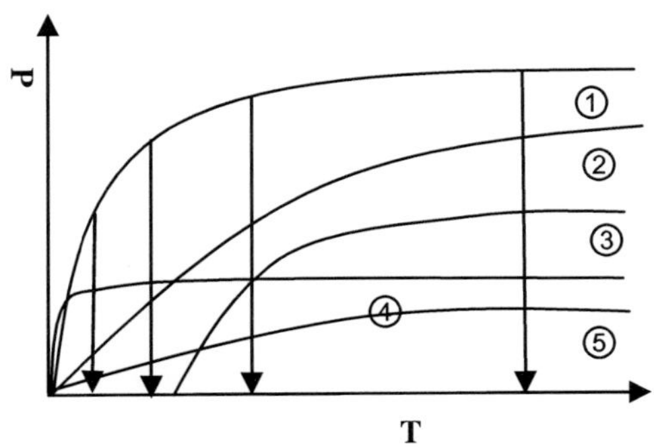

Figure 1. The time characteristics of multiple irreversible processes.

When acted under stimulus, multiple irreversible processes of a thermal system can be unwrapped and a system will show different Figures at different times. Five reversal processes are represented in this Figure. P represents the probability of one conformation involved in protein folding, which result from reversal process of protein folding.

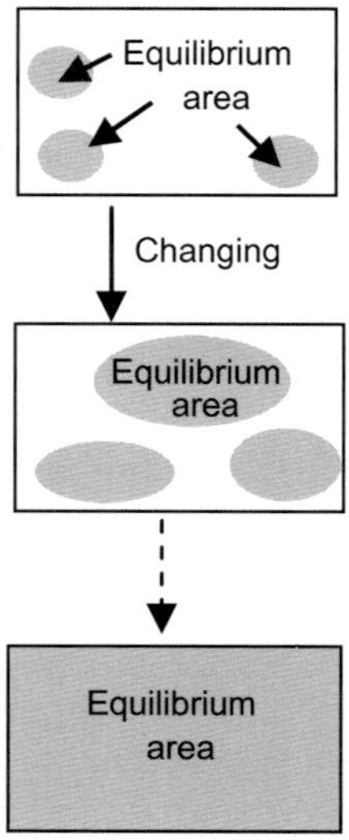

Figure 2. The time characteristics of thermal system involved in irreversible change.

When a larger system (square) is subjected to change irreversibly, many equilibrium sub-systems can be found within this system, the size of equilibrium systems is changeable with time. Finally, the whole system returns into equilibrium state and other sub-systems disappear.

Within a short time, the equilibrium state of these sub-systems can be reached in a small area (space). Within a long time, the equilibrium state of a thermal system can be reached in a larger area (space). Or in other words, the time is the control factor that determines the size (scope) of equilibrium sub-systems appearing in irreversible change of a complex system.

Let us pay attention to a system or one reversal process of thermodynamics, as it represented in Figure 3. At first, the system is at nonequilibrium state and finally transformed into equilibrium state. The transformation from nonequilibrium state to equilibrium state is totally dependent on the time the system undergoes. Thus, the property of a system is again determined by the time.

Thus, we have gotten an important conclusion: the time acts thermodynamics criterion for thermal processes.

Our analysis agrees well with traditional physics. The half-life of physical changes has been well known and is widely utilized

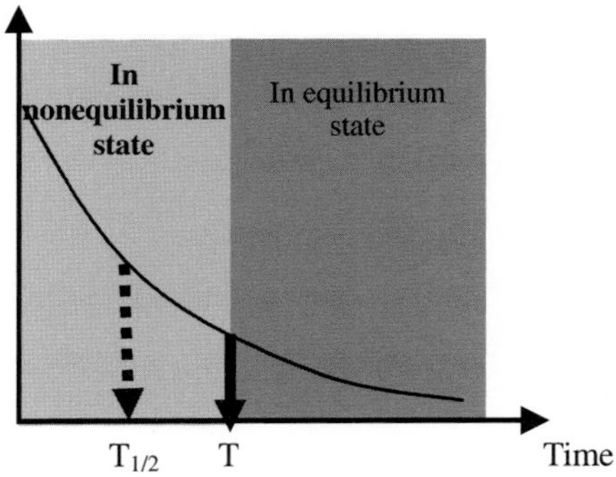

Figure 3. Transformation from nonequilibrium and equilibrium states along time.

The behavior of a system is changeable with time. We can theoretically define a point of time, between nonequilibrium and equilibrium states of a system.

in all areas of physics and chemistry, but none consider it as thermodynamic criterion for thermal process (1).

2. The Dynamics Freedom of a System (or an Reversible Process)

Among the recognition that time characterizes a thermal process and it acts thermodynamics criterion, we need to define a new scientific concept of thermodynamics: dynamic freedom of a thermal system, to quantitatively analyze the relationship between different thermal processes. The physical meaning of it is the demarcation point between equilibrium and non-equilibrium state of a thermal system. Simply speaking, within this time scale, a system, when subjected to change, can be considered at none-equilibrium state and equilibrium state beyond this point. Physically speaking, there is not a clear point, but we must arbitrarily define it. Although this dichotomy of equilibrium and none-equilibrium state represents a simplest model, it is valuable in our studying of irreversible processes of a thermal system or protein folding. See Figure 3 for explanation.

In practice, we can assume that:

$$T = 2T_{1/2}$$

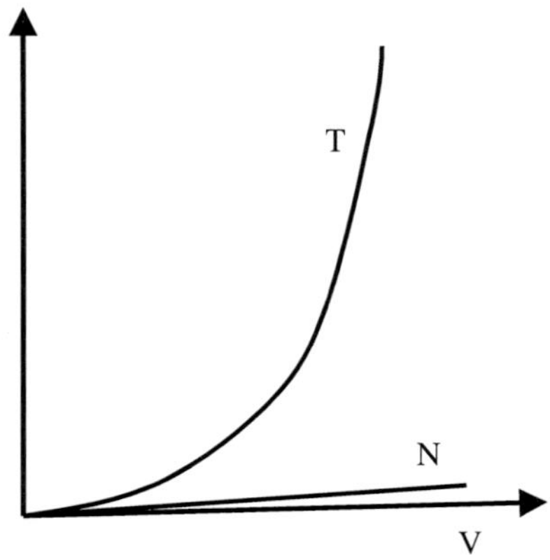

Figure 4. Relation between system volume and freedom of it.

T represents freedom of one thermal system, N is the number of particles, and V is volume of it.

Where $T_{1/2}$ represents half-life of one thermal process and T represents dynamics freedom of one thermal system. So it can be measured experimentally and conveniently.

If a thermodynamic system comprises more none-identical particles (caused by volume expansion), the freedom of this system will increase rapidly. See Figure 4. This conclusion is of importance in our analyzing of protein dynamics.

3. Protein Dynamics of Backbone of a Protein

The global protein conformational changes are induced by the rotation around backbone of a peptide or the dihedral angle of peptide plane and side chains of a protein take its effect by influencing protein dynamics of backbone. In a simple model, we can only study protein dynamics of backbone in protein folding.

In order to analyze protein folding theoretically, we need to define several basic concepts of protein dynamics.

The pother is defined as rotation around dihedral angle of peptide plane and this can induce the change of protein conformation. The potherse is defined as a thermal system of protein dynamics of backbone of a protein and it is composed by many pothers.

For a system comprising N pother (which represents the number of degree of freedom in view of thermodynamics), the state space (S or sample space of protein conformation of protein folding) is N!. The number of residues (R) is N/2. The rate of protein conformational change (υ or change of pother) is about 10^{10} turn/second.

For protein folding, we can ignore energy differences of different conformational state and we also assume that dynamics freedom of a thermal system is close to the time to search all conformational states of a system by random collision mechanism. Then:

$$T = S/\upsilon$$

$$S = N!$$

$$N = 2R$$

$$\text{If } T = 2 \text{ minutes}$$

$$N \leq 16$$

$$R \leq 8.$$

This result indicated that most stable conformation of a protein can be only obtained within size of 8 free residues (or 16 potherse). If a polypeptide is larger than this size, other mechanisms, which decrease the freedom of a system, must lie there, otherwise, the most stable state cannot be searched. (The simplest calculation indicated that error is very smaller and has no impact on our studying of protein folding, see following discussion).

By this way, the logical relationship between time scale of protein conformational change and internal motion of backbone of a protein can be established.

4. Basic Principle of Irreversible Thermodynamics Theory for Protein Folding

In 1985–1992, we hope to establish irreversible thermodynamics theory for protein folding on the basis of dissipative structure theory, developed by Prigogine (3). After 7 years studying, all our approaches had produced failure totally. In 1992, we gained a new understanding of science: it is time characteristic of protein folding, not Gibbs energy or entropy of protein folding, which determine the size or scope of diversified thermal processes of protein folding and it acts controlling factor for the origin of natural order (1, 14). On the basis of this recognition, the irreversible thermodynamics theory for protein folding and protein thermodynamics structure theory has been developed (1) on the basis of this principle.

The following principles of protein folding have been proposed: 1) protein folding is irreversible process in view of thermodynamics; 2) any irreversible process can be reconsidered as complex admixture of many types of reversible processes; 3) any reversible processes of protein folding corresponds to one subsystem of thermodynamics (one potherse or one type of protein conformational change); 4) The dynamics freedom of any reversible process of folding is smaller than the time scale of protein folding; 5) the protein folding determinants differ from protein stability determinants; 6) the restriction factor which guarantee the unique conformation (structure) of any residue of a protein within a folded protein comes from thermodynamics interaction within a protein.

The principles of our theory can be represented in Figure 5.

Figure 5a represents the folding pathways according to irreversible thermodynamics mechanism. The actual line of folding pathway can be recognized as complex admixture of many reversible processes or protein conformational change, thus the Figure 5b can be obtained. By irreversible pathway of folding, a protein can reach its most stable state of thermodynamics or heat stable state and our theory cannot tell the solution for this problem. But prion biology provides an example that native protein is not at most stable state of thermodynamics (7).

In Figure 5c, we can see that many sub-systems can be generated in irreversible processes and these systems are always at equilibrium state. For most irreversible processes of physics, the system changes with time and no steady structure can be generated. But if these systems can trap into a heat stable state by cooperation mechanism, the fashion of this division could be sustained forever; it is true for protein folding. In order to establish steady relationship between

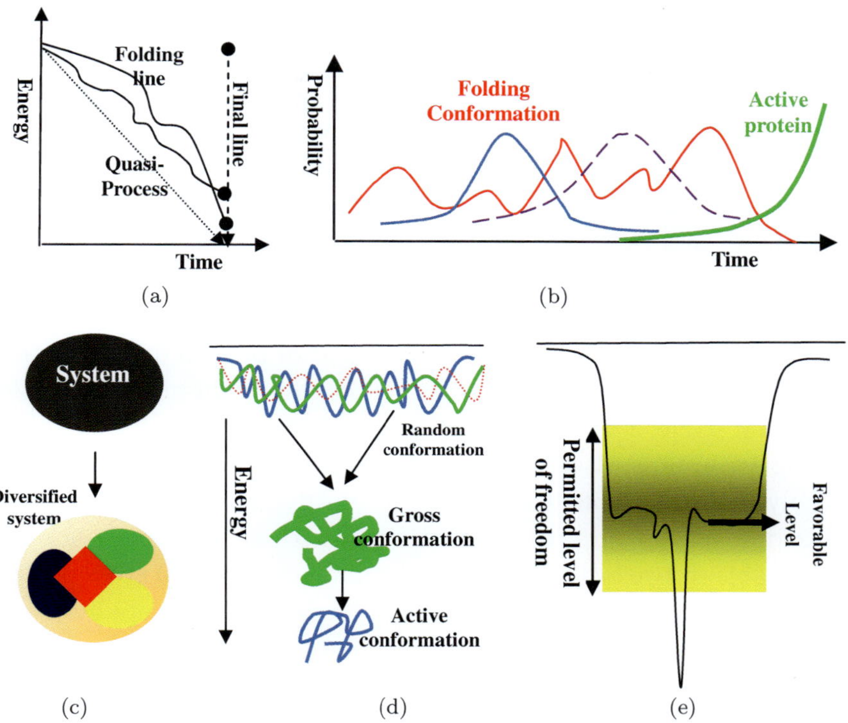

Figure 5. principles of irreversible thermodynamics theory for protein folding.

Figure 5a: the energy line of protein folding
Figure 5b: the probability of protein conformation of protein folding
Figure 5c: the division of sun-system within protein folding
Figure 5d: the sequence of diversified conformations of protein folding
Figure 5e: protein folding is accompanied with the decrease of freedom.

these systems, one system at advanced level is needed. Thus, hierarchical structure of thermodynamics (protein thermodynamics structure) is generated in folding. We have estimated that the maximum size of polypeptide in which the most stable conformation can be found by random collision mechanism is 8 free residues. The sub-system of thermodynamics is coined potherse in our theory. The cause of such division is solely based upon time scale of irreversible processes, rather than energy. This opinion is completely same with Levinthal's conclusion that a protein cannot find its most stable state of thermodynamic among giant space of possible conformations in a limited range of time by random collision mechanism (8).

The maximum size of potherse is of great importance in our theory and we can understand it by several ways. Firstly, it sets maximum length of sequence in which peptide can seek its most state of thermodynamics by random collision mechanism. An area in excess of 8 amino acid residues must have some physical factors that reduce the freedom of the thermodynamic system, or it has involved in two or more potherse, otherwise heterogeneous conformation should be generated. Secondly, it set theoretical limit of protein dynamics that freedom of all types of protein dynamics of a protein must be smaller than that value.

Figure 5d illustrates the ideal processes of protein folding. Firstly, a polypeptide adopts conformations with low stability at many sites due to short range interaction and these conformations are changeable. If two or three local conformations can form gross protein conformation by random collision mechanism, it will rapidly transform into active state of a protein by conformation reorganization. If not, the polypeptides will change its conformations forever and find next chance to form gross conformation; otherwise it will aggregate between different molecules.

In Figure 5e, the profile of freedom decrease (or energy level) of a single protein is represented. At first, the freedom of a polypeptides decreases rapidly as many local conformations are formed spontaneously. Then it changes slowly accompanied with the formation of gross conformation of a protein which is rapidly transformed into active structure of a protein. The level of freedom of peptide in the step of formation of gross conformation could be regulated by urea (15), temperature (16, 17) and other substances (18, 19). The higher freedom would decrease probability of desired conformations of folding and lower freedom would increase stability of undesired conformations and both impair the folding ability of a protein. There is most favorable level of stability of folding conformation. Protein folding stability

In views of irreversible thermodynamics theory for protein folding, diversified folding conformations (or intermediates, potherse)

co-occur spatially and temporally and this coherence (synergetic) determines folding ability of a protein. However, this prerequisite of space and time could be changeable in some degree without influencing folding ability. Thus comes concept of folding stability.

The studying on P22 protein (tailspike endorhamnosidase of Salmonella phage P22) provides good evidence for it. Some folding mutant (Temperature-sensitive folding) cannot fold at high temperature and global-tsf-suppressor mutant can recover folding ability at high temperature (16, 17).

In other aspect, the protein folding stability also provides a reasonable explanation for the multiple-mechanisms of molecular biology for protein folding. One thermodynamics pathway of folding can be carried out by many mechanistic mechanisms.

If insertion, deletion, and residues substitution do not change the folding pathway greatly and provide new gross structural fashion of a protein, it will not influence protein folding ability in essence, but will change folding rate. Many such evidences have been obtained in protein science (20, 21).

The generic profile of protein folding stability and protein stability curve (39, 40) are shown in Figure 5.

5. Prediction and Evidences

The production of protein folding is protein thermodynamics structure. This deduction is similar to that of Prigogine's theory of irreversible thermodynamics about dissipative structure (22). There is no such opinion in view of conventional view of protein science. The protein thermodynamics structure theory is supported by studying on protein dynamics and others (23).

A protein is not global system of thermodynamics, but composed of many sub-systems, or potherse. This deduction has been well confirmed by protein dynamics (24). More powerful and directly evidence comes from study of Englander. He has found that protein dynamics of backbone of a folded protein can be divided into many independent foldons. All constituents of a foldon have same m value, a method to measure local fluctuation or protein dynamics

(25–27). In my opinion, a foldon represents one potherse. In addition, one amino acid residue substitution can greatly influence the protein folding ability. This fact can be easily interpreted by our theory that it has great impact on the preference of local conformation of protein folding and thus can influence folding ability (17).

Logical separation of protein stability and protein folding. According to our theory, the protein stability is a prerequisite for protein folding and protein folding pathway is independent with protein stability under some specific cases (or models). This concept has been confirmed experimentally by the discovery of protein folding mutant that influence protein folding ability, but has no impact on protein stability (16, 17). The disulfide bond, which can great enhances stability of a protein, is not the determinants of protein folding (28). The difference between them is drawn is drawn in Figure 6.

Complex thermal interactions between different residues occur in protein folding. It has been known in protein science that contribution of residues to protein stability is related to diversified structures of a protein and their contribution cannot be simply added with each (29). The global suppression of protein folding mutant has been discovered (16, 17). But the detailed mechanism remains to be studied further.

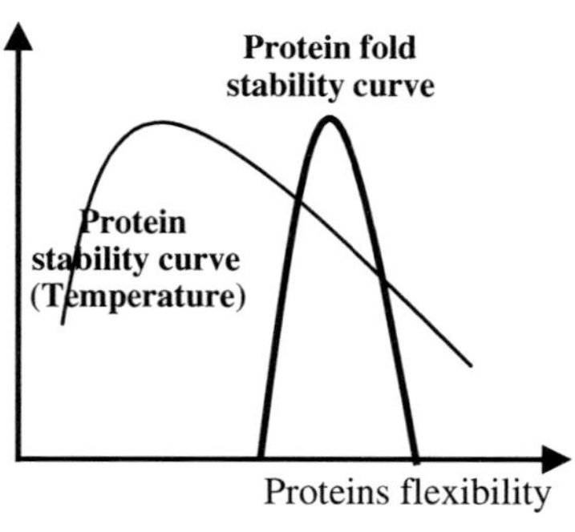

Figure 6. Generics profile of protein folding stability and protein stability.

The generic profile of protein stability and protein folding stability is shown in this Figure. We can see that protein folding determinants differs from protein stability determinants in principle.

Protein flexibility and protein folding rate. According to our theory, the protein folding is the process to decrease protein flexibility (or protein dynamics). If a folded protein shows high flexibility, the residual flexibility within folded protein will take its role in all steps of protein folding and this will slow the folding rate, and vice versa. So a protein with a tightly order structure and low flexibility will fold faster (30, 31).

Linker (or loop area) of a protein. For the linker (loop) area between two secondary structures of a protein (5, 32, 33), see Figure 7, our theory believes that the homogeneous conformation in this area is guaranteed by it's intrinsically structure both in protein folding and final structure. Within this area, a long sequence excess 8 residues, such as $(Gly)_8$, without any structural restriction will not be permitted. Otherwise, homogeneous conformation in this area cannot be guaranteed theoretically. Normally, the loop area of a protein shows high protein flexibility and it has great impact on the folding rate of a protein. This prediction has been examined by Weikl (5).

The principles of rules 6 may be invalidated under specific case. The correctness of principle 6 is highly questionable. However, without it, we can not calculate the size (range) of a potherse. It seems that concept of unique conformation of any residues of a protein is not true under some cases albeit concept of unique structure of a protein is widely accepted. Our theory cannot tell

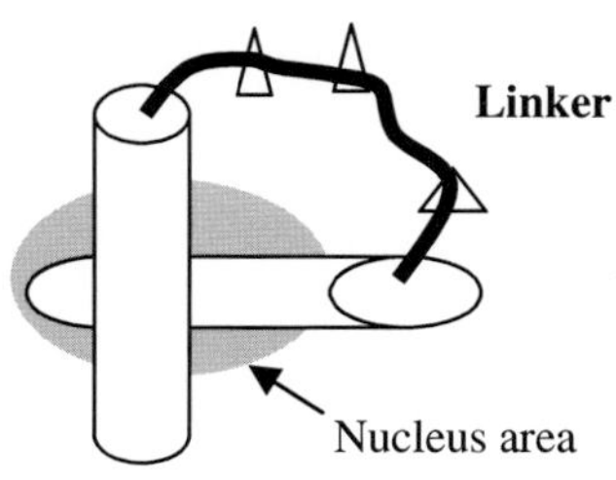

Figure 7. Linker of secondary structure.

$^{\triangle}$) sites which controls rate of protein folding. The cylinders represent secondary structure of the protein. If there are no structural determinants in linker area, it can fold according to mechanism of equilibrium thermodynamics and show heterogonous conformation in this area.

the actual truth. Whatever the case is, our theory provided a strict, ideal and theoretical model for protein folding and revision should be made under specific conditions.

The logical separation of protein sequence and protein folding. According to our theory, the bioinformation of protein sequence first transform into thermodynamics factors, which lower the protein mobility, and then can be integrated within diversified steps of protein folding. In one aspect, the distribution of protein flexibility along protein sequence is related to protein folding pattern. Thus, many different proteins with different sequences, can fold into similar shape (34, 35).

The protein folding is the integration process of bioinformation. According to irreversible thermodynamics theory for protein folding, the information stored in protein sequence firstly transform into thermal factors, and then can be integrated in protein folding and finally result in the active structure of a protein. The thermal interaction between different factors can be neutralized, cooperated, added, interfered, annexed, divided and etc. at diversified systems at diversified levels (or secondary, tertiary structures of a protein). The information of protein folding is much bigger than information required for protein stability or three-dimensional structure. Much information of protein folding has no contribution to protein stability. So, the folding cannot be reconsidered as translation process by any way. Or in other words, there is no unified protein folding codes (36, 37).

Irreversible thermodynamics theory shows powerful ability to integrate experimental models: As a pure thermodynamics and theoretical model of protein folding, the irreversible thermodynamics theory cannot interpret experimental data of protein folding directly. The extra-hypothesis must be firstly introduced and it can then gain such ability, or in other words, the experimental model must be established based upon it and this experimental model have such ability to explain experimental data. In fact, many experimental models have been proposed and the rightness of these models can be demonstrated logically by irreversible thermodynamics theory for protein folding.

Scheraga's theory proposed in 1971 can be fully interpreted theoretically by our theory (1, 23). In fact, our theory is the theoretical explanation for his theory on the basis of irreversible thermodynamics. According to our theory, the short range interaction and long range interaction represent different reversible processes of protein folding which are tightly related (cooperated); the short range interaction controls the probability of folding conformation at local site (similar to secondary structure of a protein, but cannot be completely same), the long range interaction controls formation of gross conformation of a protein. The formation of local conformation governs the spatial relation between different parts of polypeptides and thus influences the formation of gross conformation of a protein. As a specific type of local structure, β-bend of polypeptides has great impact on the folding ability of a protein for which control the topological structure of polypeptide (38, 39). The hydrophobic interaction (long range interaction) may pay an important role in the formation of gross conformation (40). But the hydrophobic interaction cannot act sole factor in controlling the formation of gross conformation of a protein; otherwise the heterogeneous conformation should be generated in other areas of a protein. In order to guarantee protein folding processed orderly, the formation of local conformation and gross conformation must be co-occur.

If cooperation of the formation of local conformation and gross conformation is stronger, the two states of protein folding (41, 42) can be observed experimentally (see Figure 5E for detail). This phenomenon is most obvious in the folding of smaller protein. For larger proteins, the formation of local conformation and gross conformation differs from each other obviously (43).

6. The Protein Folding and Unfolding were Processed by Different Ways

The critical evidence for the irreversible thermodynamic theory for protein folding is the differences between folding and unfolding test.

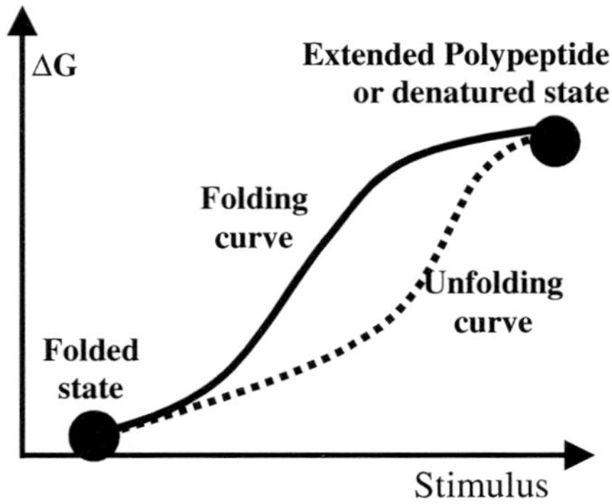

Figure 8. Diagram of folding and unfolding curves.

The folding curve differs from that of unfolding. The unfolding curve is getting close to equilibrium curve.

There are many facts that indicated that protein folding is not the inverse process of unfolding. Here we summarize it as follows.

1. The protein mutant which affect its stability but no folding ability have been discovered.
2. The s-s bond can great change protein stability, but it has little or nothing to protein folding.
3. Normally, the protein folding is processed slowing than that
4. The protein folding cannot process under some environments and in these environments a protein shows high stability.

Thus, the unfolding is not the inverse process of folding. Figure 8 illustrates the differences between two processes.

7. Multiple Mechanisms for Protein Folding

Physically speaking, there are many mechanisms for protein folding. According to current evidences obtained, it is reasonable to conclude that native protein fold by way of irreversible thermodynamic. However, other ways of protein folding are possible and may be true for some specific sequences of protein under specific conditions.

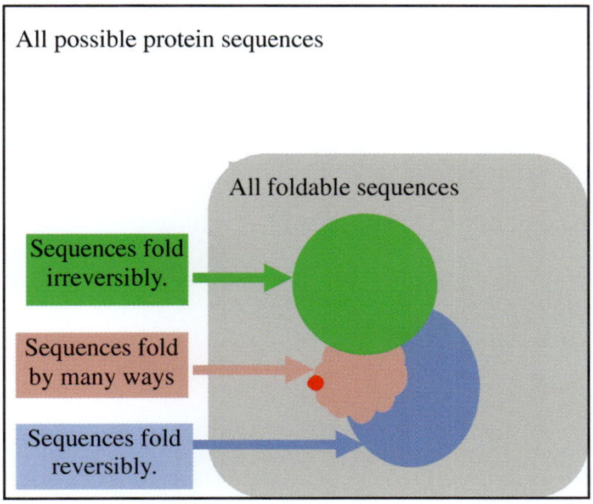

Figure 9. The relationship of different mechanisms.

Different sequences may adopt different ways of folding. This Figure is only my guess and the experimental studying about it is still lacking.

The relationship between these mechanisms is diagrammed in Figure 9.

1) Irreversible thermodynamic mechanism was true for native protein. The physical rules cannot guarantee that all possible sequences have power to fold into uniform protein structure. Our opinion is that sequence that cannot fold into structured protein has been eliminated in biological evolution.
2) Equilibrium thermodynamic mechanism may occur for some sequence, but it is untested in protein science. But DNA science has provided good example. The denatured DNA strand renature into duplex DNA under favorable condition by ways of Equilibrium thermodynamic mechanism. In contrast, denatured DNA fold into coil structure under unfavorable condition by way of irreversible thermodynamics.
3) It is possible to find some sequences which can fold into many types of protein structure. The prion can be considered as one example (7).

4) Some proteins may fold into kinetics stable structure, rather than thermodynamics stable structure. According to our theory, a protein may fold into heat stable state and it may be unable in views of thermodynamics. The prion biology has provided a powerful evidence for it. In addition, it has found that many proteins are kinetics stable (heat stable), rather than thermodynamics stable (6, 9, 10).

In addition, some sequences are unstructured in vivo (44, 45).

References

1. Zhao Q (2001) Irreversible thermodynamics theory for protein folding and protein thermodynamics structure. *Prog Biochem Biophys* **28**, 429–435.
2. Privalov PL (1989) Thermodynamic problems of protein structure. *Annu Rev Biophys Biophys Chem* **18**, 47–69.
3. Prigogine I (1980) *From Being to Becoming*: Time and Complexity in the Physical Sciences (Freeman) San Francisco.
4. Anfinsen CB (1973) Principles that govern the folding of protein chains. *Science* **181**, 223–230.
5. Dill KA, Ozkan SB, Shell MS and Weikl TR (2008) The protein folding problem. *Annu Rev Biophys* **37**, 289–316.
6. Nelson CA and Hummel JP (1962) Reversible denaturation of pancreatic ribonuclease by urea. *J Biol Chem* **237**, 1567–74.
7. Prusiner SB (1991) Molecular biology of prion diseases. *Science* **252**, 1515–1522.
8. Levinthal C (1968) Are there pathways for protein folding? *J Chim Phys* **65**, 44–45.
9. Baker D, Sohl JL and Agard DA (1992) A protein-folding reaction under kinetic control. *Nature* **356**, 263–5.
10. Baker D and Agard DA (1994) Kinetics versus thermodynamics in protein folding. *Biochemistry* **33**, 7505–9.
11. Popov EM (1999) Protein folding as a nonlinear nonequilibrium thermodynamic process. *Biochem Mol Biol Int* **47**, 443–53.
12. Hartl FU (1996) Molecular chaperones in cellular protein folding. *Nature* **381**, 571–9.
13. Vendruscolo M, Zurdo J, MacPhee CE and Dobson CM, (2003) Protein folding and misfolding: a paradigm of self-assembly and regulation in complex biological systems. *Philos Transact A Math Phys Eng Sci* **361**, 1205–22.

14. Schrödinger E (2002) What Is Life? Cambridge: Cambridge University Press.
15. Nichtl A, Buchner J, Jaenicke R, Rudolph R and Scheibe T (1998) Folding and association of β-galactosidase. *J Mol Biol* **282**, 1083–1091.
16. Aramli LA and Teschke CM (1999) Single amino acid substitutions globally suppress the folding defects of temperature-sensitive folding mutants of phage P22 coat protein. *J Biol Chem* **274**, 22217–24.
17. Mitraki A, Fane B, Haase-Pettingell C, Sturtevant J and King J. (1991) Global suppression of protein folding defects and inclusion body formation. *Science* **253**, 54–58.
18. Cleland JL and Wang DC (1990) Cosolvent Assisted Protein Refolding. *Nature Biotechnol* **8**, 1274–1278.
19. Wetlaufer DB and Xie Y (1995) Control of aggregation in protein refolding: a variety of surfactants promote renaturation of carbonic anhydrase II. *Protein Sci* **4**, 1535–1543.
20. Heinz DW, Baase WA, Dahlquist FW and Matthews BW (1993) How amino acid insertions are allowed in an α-helix of T4 lysozyme. *Nature* **361**, 561–564.
21. Weikl TR (2008) Loop-closure principles in protein folding. *Arch Biochem Biophys* **469**, 67–75.
22. Kondepudi D and Prigogine I (1998) Modern Thermodynamics: from heat engine to dissipative structure. New York. John Wiley Press.
23. Zhao Q (2009) Protein thermodynamic structure. *IUBMB Life* **61**, 600–606.
24. Zhao Q (2011) dynamic model of enzymes action. *Protein Pept Lett* **18**, 92–99.
25. Sabrina Bédard, Leland Mayne, Ronald W Peterson, A Joshua Wand and S Walter Englander (2007) The foldon substructure of staphylococcal nuclease. *Journal of Molecular Biology* **376**, 1142–1154.
26. Haripada Maity, Mita Maity and S Walter Englander (2004). How cytochrome c folds, and why: submolecular foldon units and their stepwise sequential stabilization. *Journal of Molecular Biology* **343**, 223–233.
27. Yawen Bai, Tobin R Sosnick, Leland Mayne and S Walter Englander (1995). Protein folding intermediates: Native-state hydrogen exchange. *Science* **269**, 192–197.
28. Wetzel R, Perry LJ, Baase WA and Becktel WJ (1988) Disul-fide bonds and thermal stability in T4 lysozyme. *Proc Natl Acad Sci USA* **85**, 401–405.
29. Tokuriki N, Stricher F, Schymkowitz J, Serrano L and Tawfik DS (2007) The stability effects of protein mutations appear to be universally distributed. *J Mol Biol* **369**, 1318–1332.

30. Bai Yawen, Zhou H and Zhou Y (2004) Critical nucleation in the folding of small apparently two-state proteins. *Protein Sci* **13**, 1173–1181.
31. Plaxco KW, Simons KT and Baker D (1998) Contact order, transition state placement and the refolding rates of single domain proteins. *J Mol Biol* **277**, 985–94.
32. George RA and Heringa J (2002) An analysis of protein domain linkers: their classification and role in protein folding, *Protein Eng* **15**, 871–879.
33. Robinson CR and Sauer RT (1998) Optimizing the stability of single-chain proteins by linker length and composition mutagenesis. *Proc Natl Acad Sci USA* **95**, 5929–34.
34. Orengo CA, Jones DT and Thornton JM (1994) *Protein superfamilies and domain superfolds. Nature* **372**, 631–634.
35. Ye Y and Godzik A (2003) Flexible structure alignment by chaining aligned fragment pairs allowing twists. Bioinformatics (Supplement 2): II246–II255.
36. Tejedor JR and Valcárcel J (2010) Gene regulation: Breaking the second genetic code. *Nature* **465**, 45–6.
37. Zhao Q (2011) Nonalgorithmicity and algorithmicity of protein science. *Advance in Bioscience and Biotechenology* **2**, 340–346.
38. Lewis PN, Momany FA and Scheraga HA (1971) Folding of polypeptide chains in proteins: A proposed mechanism for folding, *Proc Natl Acad Sci USA* **68**, 2293–2297.
39. Scheraga HA, Wright PE and Dyson HJ (2006) Role of specific hydrophobic regions in initiating protein folding. *Proc Natl Acad Sci USA* **103**, 13057–13061.
40. Hao MH and Scheraga HA (1998) Theory of two — state cooperative folding of proteins, *Acct Chem Res* **31**, 433–440.
41. Chen RPY, Huang JJT, Chen HL, Jan H, Velusamy M, Lee CT, Fann W, Larsen RW and Chan SI (2004) Measuring the refolding of {beta}-sheets with different turn sequences on a nanosecond time scale. *Proc Natl Acad Sci USA* **101**, 7305–7310.
42. Miranker AD and Dobson CM (1996) Collapse and cooperativity in protein folding *Curr Opin Struct Biol* **6**, 31–42.
43. Ivankov DN and Finkelstein AV (2004) Prediction of protein folding rates from the amino acid sequence-predicted secondary structure. *Proc Natl Acad Sci USA* **101**, 8942–4.
44. Dyson HJ and Wright PE (2005) Intrinsically unstructured proteins and their functions. *Nat Rev Mol Cell Biol* **6**, 197–208.
45. Uversky, VN (2002) Natively unfolded proteins: A point where biology waits for physics. *Protein Sci* **11**(4), 739–756.

Chapter 11

Can We Go from Protein Sequence to Protein Structure?

Summary: A protein has its unique three-dimensional structure and sequence. In protein science, a well-accepted view is that information of protein three dimensional structure and protein folding has been prescribed within sequence of a protein. Thus, some scientists believe that we can find a program, on the basis of which the protein three dimensional structure could be deduced from protein sequence. But the solution for it has not been obtained now. Our conclusion is that we cannot predict all protein three-dimensional structures from protein sequences using one program, nor can we predict protein function from protein three dimensional structure with one program. This represents the logic limit of protein science.

Globular proteins have its unique three dimensional structure and sequence. So there is a fundamental relationship between protein sequence and its function. Anfinsen has proposed that protein sequences have all information for protein three dimensional structure and folding within model of equilibrium thermodynamic theory for protein folding. Combination with dogma of molecular biology, the second codes of genetics (or protein folding codes) has been proposed. It states that we can figure out protein three dimensional structure from protein sequence and exposition of such principles are the central task of molecular biology. In fact, many scientists have devoted themselves to this topic of science and remarkable achievement has been obtained. But we are far away from doing so now. Great difficulties which we have encountered enforce us to ask if it is possible to predict protein structure from protein sequence. If the answer is no, the biology would change from current path to another, and we should recheck our thought about biology at fundamental level.

Our theory for protein folding shows that protein folding is a typical process for the origin of natural order (1–6). The hierarchical structure of protein three-dimensional structure is formed and information of protein folding is integrated at diversified levels within protein hierarchical structure expressed in protein folding. In this paper, we will discuss logic principles (metaphysical properties) of protein folding. Our conclusion is that we cannot predict all protein three dimensional structures from protein sequences using only one program, nor can we predict protein functions from protein three dimensional structure with only one program. This represents the logic limit or theoretical limit of protein science (7). In this chapter, we demonstrate this conclusion and some relevant questions on protein folding. First, the conclusions were discussed and then reader can have a clear thought about it. Next, some evidence and related matters are discussed.

1. There is No Fixed Way by Which We Can Go from Protein Sequences to Structures

As we have learnt from irreversible thermodynamic theory for protein folding, a protein is a complex system of thermodynamics and it has infrastructure. The protein three dimensional structure only represents a static figure of a protein expressed in crystal state, and not in solution. In solution, a protein changes its shape for ever and there is no fixed three dimensional structure, but all types of protein conformational change of native protein occur in a range which cannot destroy its unique structure. In view of protein thermodynamic structure theory, a protein comprises many thermal systems and thus the relationship between protein structure and protein sequence (or its constituents) can be only described by system theory. By applying principles of system logic into field of protein science, the following conclusions could be obtained:

1) For a folded protein, there is at least one program that allows prediction of protein three-dimensional structure from the protein sequence.
2) No program exists for predicting all protein three-dimensional structures from protein sequences.
3) We must predict different proteins (maybe protein structure families) by different programs.

These conclusions are also validated in the structural genome (2, 3). We can revise this as follows:

1) For a protein function (or property), there is at least one program for finding the connection between it and the protein structure (conformation).
2) No existing program can predict all protein functions from protein three-dimensional structures. In other words, the models that describe the relationship between protein function and conformation are incompatible with each other.
3) We must predict different protein functions by different programs.

We must point out that these conclusions do not mean that the protein structural prediction is nonsense. Our opinion is that a program that could predict that some types of protein structure is valuable and that pursuing a more precise and absolute right theory for protein structure prediction is unnecessary.

Consistent with our conclusion, the work of Dobbins *et al.* (8) has shown that a composite model is necessary to describe the diversity of conformational change observed during molecular recognition.

2. The Cooperation Phenomena of Protein Conformational Change

The cooperation of protein conformational change is universal both in a folded protein and unfolded protein (protein folding process). Its logic structure could be expressed as follows.

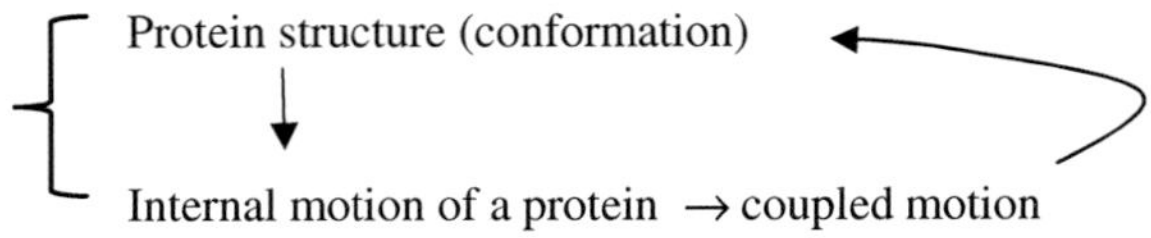

In view of experimental science, the logic cycle between protein secondary structure and tertiary structure is well established. For a folded protein, there is interaction between protein secondary structure and tertiary structure. Normally, protein tertiary structure regulated the fashion of secondary structure and vice versa. For example, a sequence of peptide will show different conformations when it embedded into different proteins.

Theoretically, the formation of protein secondary structure is also processed cooperatively. All secondary structures, helix, turn, are composed of residues which are tightly adjacent in protein sequence.

These analyses show that system logic theory is validated in protein science.

3. Protein May Not be at Its Most Stable State *In Vivo*

The belief that protein structure can be predicted from protein sequence comes from hypothesis of Anfinsen; it states that protein is at its most stable state of thermodynamics in living environment. If so, we can certainly go from the sequence to the structure. At least, we can calculate all possible conformations and the lowest conformation will be the native state of a protein. But this hypothesis is not compatible with many evidences. Here we cite some evidences that cannot be explained by Anfinsen's hypothesis.

1) Prion biology has provided a good example that native state of a protein is not at most stable state *in vivo* (9).
2) By gene manipulation, it is very easily to fuse two genes into one. By this way, many fusion proteins, two protein sequence are linked up into a whole, has been expressed in vivo. For these fusion polypeptides, different parts (or proteins) are

folding independently. According to Anfinsen's hypothesis, these polypeptides should find their most states in whole sequence, not in its original sequences.

3) The thermodynamic stability of proteins is rather low, most of which are about 5–10 kcal/Mol. This indicates that protein stability in solution comes from its dynamic stability. Or in other words, the activation energy of protein denaturing is very high. In some cases, the thermodynamic stability of a protein is near zero, and the protein still shows high dynamic stability. Thus the thermodynamic stability of a protein is not the controlling factor for protein conformational change as it did in most chemical reactions.
4) Differences between protein stability determinants and folding determinants (10). It shows that protein folding determinants has nothing to do with protein thermodynamic stability. Thus, if the role of folding residues has been considered in a program to predict protein structures, it is questionable. We should not ignore the role of protein folding determinants.

4. Protein Folding is the Integrating Process of Bio-Information

One way of experimental science to demonstrate the predictability of protein structure from protein sequence is whether the protein folding is an information integrating process. If it is, it could be integrated by many ways (or programs in terms of mathematics), we cannot go from the protein sequence to the structure by a fixed way.

Many evidences confirm that a protein folding is an integrating process of information stored in protein sequence and environmental factors.

Firstly, the cooperation, neutralization (global suppression) phenomena of effect of residues on protein folding has been observed (10).

Secondly, protein folding can be processed under some conditions and the interaction between information of protein sequence and environmental factors occurs (10).

Thirdly, the protein folding unit of thermodynamics, or the folding concept has been proposed (11).

Fourthly, different protein sequences can fold into similar structures.

Most importantly, our model can give a satisfying explanation to many experimental results located in various fields of science.

References

1. Anfinsen, C. B. (1973) Principles that govern the folding of protein chains. *Science*, 181, 223–230.
2. Watson, J. D., Laskowski, R. A. and Thornton, J. M. (2005) Predicting protein function from sequence and structural data. *Current Opinion in Structural Biology*, 15, 275–284.
3. Jones, D.J., Sternberg, M. J. E. and Thornton, J. M. (2006) Bioinformatics: From molecules to systems. Philosophical Transactions of the Royal Society B: *Biological Sciences*, 361, 389–391.
4. Onuchic, J. N., Luthey-Schulten, Z. and Wolynes, P. G. (1997) Theory of protein folding: The energy landscape perspective. *Annual Review of Physical Chemistry*, 48, 545–600.
5. Zhao, Q. (2001) Irreversible thermodynamics theory for protein folding and protein thermodynamics structure. *Progress in Biochemistry and Biophysics*, 28, 429–435.
6. Zhao, Q. (2011) Nonalgorithmicity and algorithmicity of protein *science. Adv. Biosci. Biotechnol.* 2, 340–346.
7. Barrow, J. D. (1998) Impossibility: The limits of science and the science of limit. Oxford University Press, London.
8. Dobbins, S. E., Lesk, V. I. and Sternberg, M. J. E. (2008) Insight into protein flexibility: The relationship between normal modes and conformational change upon proteinprotein docking. *Proceedings of the National Academy of Sciences of USA*, 105, 10390–10395.
9. Collinge, J., Sidle, K. C., Meads, J., Ironside, J. and Hill, A. F. (1996) Molecular analysis of prion strain variation and the aetiology of "new variant" CJD. *Nature*, 383, 685–690.
10. Aramli, L. A. and Teschke, C. M. (1999) Single amino acid substitutions globally suppress the folding defects of temperature-sensitive folding mutants of phage P22 coat protein. *Journal of Biological Chemistry*, 274, 22217–22224.
11. Englander, S. W. and Mayne, L. (2014) The nature of protein folding pathways. *Proceedings of the National Academy of Sciences*, 111(45), 15873–15880.

Chapter 12

Protein Thermodynamic Structure Theory

Summary: A protein is not a uniform system of thermodynamics, but is composed of many sub-systems of thermodynamics, called potherse. This sub-system can be structured differently, thus comes the concept of protein thermodynamic structure. In the following chapters of this book, we will show that properties of a protein can be analyzed by applying principle of protein thermodynamic structure theory. Then this chapter will discuss basic principles of protein thermodynamic structure theory.

The current knowledge of protein science arises from its treatment as an experimental science. Because in field there are many different models, which lack a coherent essential relationship, many fundamental questions of protein science remain unresolved, and a fundamental theory of protein science is urgently required. The key task of theoretical study of protein science is to develop an axiomatic theory and a deductive system that incorporates current knowledge of protein biochemistry. Protein thermodynamic structure theory has provided such theoretical framework. In this chapter, we will discuss basic principle of it.

As we have learnt in irreversible thermodynamic theory for protein folding, many sub-systems of thermodynamics (potherse) have been formed in steps of protein folding (1, 2). When the folding of a protein is finished, these sub-systems are integrated into a complex system of thermodynamics and protein three-dimensional structure is finally formed at final step of protein folding. In a folded protein, a protein is still not a uniform system of thermodynamics and many sub-systems (potherse) can be isolated both theoretically and experimentally. The relation among these sub-systems could be described by protein thermodynamic structure theory.

1. A Protein is Not the Uniform of a Thermal System

The theoretical foundation of protein thermodynamic structure theory is that a protein is not a uniform system of thermodynamic as it did in common materials of solid. Thus, we should study its property it carefully.

When a protein acts on a new environment, such as pH, salt concentration, denaturants, its global structure (or global conformation) remains unaltered, but some local conformations could be disturbed (3). For example, when urea was added into the working buffer of enzyme, the enzyme activity has lost, but the protein structure remains unchanged; when the urea was removed from working buffer, the enzyme shows its activity again. If more urea was added into working buffer of the enzyme, the protein structure of enzyme will be destroyed and enzyme was denatured. In this case, when urea was removed from the buffer, the enzyme activity will not be recovered. From this well-established fact, we have learnt that the separation of global and local conformation of a protein. In other words, the local and global conformation show independence in some degree, the disturbing of one type of conformation may be not disturbing other conformation. Thus, a protein is not a uniform system of thermodynamics. It is a powerful evidence for protein thermodynamic structure theory.

The protein-ligand interaction has provided another evidence for it. A protein can bind to many types of substance, such as regulator and ligand. It has shown that the binding affinity between protein and regulator may not be influenced by other type of protein ligand interaction.

It is well known that the gene mutation (Insertion, deletion, and residue substitution) of a protein can alter the local conformation of a protein at one site, but in most cases, the whole or global structure of the protein remains unchanged. Thus, the modification of local conformation of a protein is independent to global protein of the protein.

Even in view of protein three-dimensional structure theory, the independence of local structure of a protein has been well documented. According to current views of protein science, a protein shows primary, secondary, tertiary, quandary structures. Same tertiary structure of proteins may be established by different secondary structure. It indicates the independence of the secondary or tertiary structure of a protein. But most scientists do not understand in view of protein dynamic and thermodynamics.

Recently, the foldon concept has been proposed. The cooperation and independence of internal motion of a protein can be clearly seen in results from such studying (4). A foldon represent a unit of cooperated internal motions within a protein and it acts a unit for protein conformational change. In my opinion, a foldon represents one specific potherse and its properties agrees with prediction of our theory completely.

Thus, the conclusion that a protein is not a uniform system of thermodynamics is beyond doubt.

2. Isolating a Sub-System of Thermodynamics within a Protein

By applying principles of system logic we have discussed in previous chapters, following conclusion could be obtained.

For any measurement or a property of a protein, a sub-system of thermodynamics can be isolated from complex thermal system of a protein and it represents the behaviors for the desired property of a protein.

For example, when a ligand binds to a receptor, a ligand binding sub-system (potherse) could be isolated in complex thermal system of the receptor. In terms of biochemistry, it refers to ligand binding conformation of a receptor. The property of interaction between ligand and a protein could be represented by behaviors of this subsystem.

In enzymology, the enzyme activity is a special measuring method for enzyme; it represents a special sub-system of thermodynamics. In biochemistry, it is called active conformation of enzyme. As conformation concept cannot describe dynamic nature of a protein, the precise terms should be an active potherse of enzyme. Within this potherse, three other types of potherse can be further isolated; they are substrate binding potherse (labeled with parameter of K_m), product binding potherse (labeled with parameter of K_p), transition potherse of enzymatic reaction (labeled with parameter of activation energy of enzymatic reaction).

Many types of protein conformations of receptor have been isolated, we have summarized it in Table 1.

Table 1. Protein conformations of receptors have been isolated experimentally.

Potherse	Current terms	Property labeling (or isolating) it
Active potherse (5) (whole protein concept)	Active receptor conformation	Activity of a receptor (whole Protein structure)
Inactive potherse (5) (whole protein concept)	Inactive receptor conformation	Inactivity of a receptor (whole Protein structure)
Desensitized potherse (5) (whole protein concept)	Desensitized receptor conformation	Inactivity of a receptor and it cannot transform itself into active conformation reversibly.
Agonists binding potherse	Ligand binding conformation	Ligand binding ability
Active Agonists binding potherse	None	Potency of agonist, ligand binding ability
Antagonist binding potherse	Antagonists binding conformation	Activity, binding ability
Regulating potherse	Regulation conformation	Regulation of receptor activity
Coupling potherse	None	Loss of protein allosteric regulation
G protein binding potherse	G protein binding conformation	Ability binding to G protein
Effector potherse	None	Essential part for receptor activity

3. Scientific Terms for Sub-System of Thermodynamics

At present, there are three different concepts that have been utilized to describe properties of a protein. They are protein conformation, protein conformational state, and potherse.

The protein conformation concept was widely utilized in field of biochemistry, molecular biology by experimental scientists. It takes advantage that this concept is compatible with structural biology and protein three-dimensional structure theory. Theoretically speaking, it cannot describe a thermodynamic state of a protein and it is unable to analyze properties of a protein. But in fact, this term has widely utilized in field of thermodynamics and thus confusion has been produced. For example, the active conformation of a protein or enzyme is a common language in our studying of protein denaturation. The underlying reason is that protein conformational change can be normally observed experimentally in these cases. But in fact, it represents a thermodynamic state of a protein for it can undergo change within it. This feature could be judged by a change of ligand-protein binding affinity, the Km of enzymatic reaction, or the protein stability in a changeable environment.

The protein conformational state is used by biophysicists who study protein thermodynamics or dynamics. It is a term of experimental science because that the logic relationship between protein conformation and thermal state of a protein cannot be interpreted within this theory.

The terms of potherse was defined in irreversible thermodynamic theory for protein folding. The pother represents a basic unit of internal and thermal motion of a protein and potherse represents a sub-system of thermodynamics which is composed of by many pothers. It takes advantage that logic relationship between internal motion of a protein and sub-system of thermodynamics of a protein could be established theoretically.

The infrastructure of a potherse is expressed as follows.

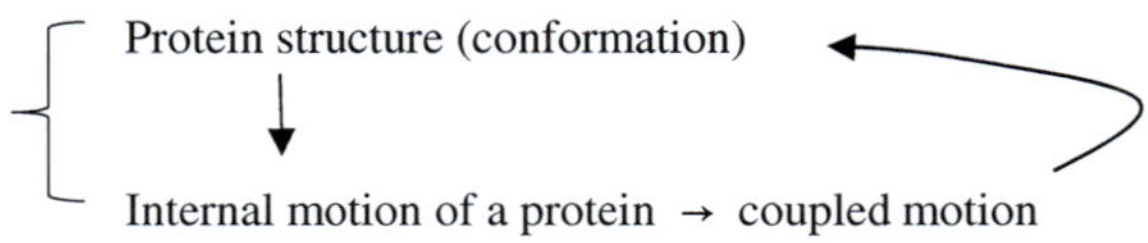

We can see that a potherse is always related to one type of protein conformation (structure). Thus, we can use conformation to describe a potherse or properties of that potherse under some cases.

One problem is that a potherse may comprise many internal motions of residues which are far away from each other both in protein sequence or space. See Figure 1 for illustration.

In this case, there is no precise concept of protein conformation within this potherse, its acts as a philosophical concept.

Finally, we have summarized the consensus and difference of these concepts in Table 2.

4. The Size and Scope of a Potherse

The maximum value of the freedom of a potherse is eight free amino acid residues. Taking into consideration that the thermal motion of residues is not free in a folded protein, the actual value of residues

Figure 1. Constituents of a potherse.

A potherse is composed of thermal motions of many residues (red triangles) which are not adjacent with each both in protein sequence and space.

Table 2. Consensus and differences in different protein concepts.

	Consensus and differences in different protein concepts		
	Potherse	Protein conformation	Protein conformational state
Axiomatic theory	√	X	X
Logic independence	√	X	X
Whole protein concept	X	√[(a)]	√[(a)]
Scientific subject	Motion	3D structure	Particle
Theoretical foundation	**Irreversible thermodynamics**	3D structure	Equilibrium thermodynamics
Model	Protein thermodynamics Structure theory	Many experimental models	Landscape theory experimental science
Eigenvalue	Time	3D structure	ΔG
Number	Infinite	Much	Infinite

(a): It originally describes whole property of a protein, but gradually it can refer to a part of a protein, but the division of protein cannot be demonstrated theoretically and logically within this theory.

may be up to 32 or perhaps slightly more. In fact, most secondary structure is smaller than that size.

In my opinion, a foldon represents specific potherse and its size is same with that we have predicted (4).

Although a potherse comprises a limited number of residues, these pothers (internal motion of a protein) can exist at different sites of a protein, and thus the scope of a potherse can cover all areas of a protein. See Figure 1 for illustration.

5. Relationship Between Different Potherses

As there are infinite types of protein conformations within a protein, there is no fixed Figure for protein thermodynamic structures, as well as protein three dimensional structure. Under a specific condition the problem could be simplified and a few types of protein conformation can be considered. Here we summarized some typical fashions of protein thermodynamic structure.

Firstly, two potherse are independent with each other completely. Oligomeric enzyme which are composed of many subunit, and each active site of enzyme activity is related to one potherse. If this enzyme shows no cooperative behavior, these potherses are completely independent.

Second, two potherse are overlapped or coupling. For example, the regulator binding conformation and agonist binding conformation of a receptor is coupled and thus the receptor activity can be regulated.

Third, one potherse may comprise other potherses. It represents one type of hierarchical structure.

The hierarchical structure of a protein thermodynamic structure can be clearly seen in protein denaturation test. We summarized it in Figure 2.

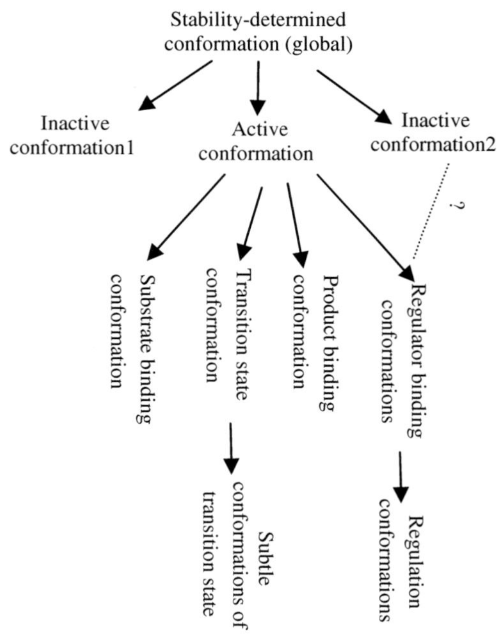

Figure 2. General fashion of hierarchical structures of protein conformational states (potherse) of enzymes.

The hierarchical structure of some types of conformations (or potherse) of enzymes is shown in this Figure.

6. The Relation of Different Potherses is Changeable

A highlighted nature of protein thermodynamic structure is that the fashion of protein thermodynamic structure is changeable, the relation among potherses within a protein is also changeable under different condition.

The first example is that cooperative nature of enzymatic kinetics is changeable. Or in other words, an enzyme can show negative cooperative in one condition and show positive cooperative in another condition. In this case, an enzyme has two or many active sites of enzymatic reaction. Each active site corresponds to one potherse which relates enzyme activity. The change of nature of cooperative phenomenon was produced from reorganization of protein thermodynamic structure.

Another example is the regulation of protein activity. The potency of agonist is changeable under different conditions. This indicates that the coupling between potherse of agonist binding and potherse of effector is changeable (6).

There is debate about the relation between enzyme activity and protein stability. Normally, enzyme isolated from thermophile is more stable and its optimum activity can be only exercised at high temperature; in contrast, enzyme isolated from mesophile has law ability. A hypothesis is that there is general relation between protein stability and optimum temperature of enzyme (7). The high stability of protein will increase its optimum temperature of enzyme. But protein with high stability and high activity at low temperature is also reported. Our opinion is that the relationship between protein stability and enzyme activity is changeable. The correlation between them expressed in behavior of native enzymes is the result of biological selection, this relation cannot be integrated by applying laws of physics.

7. The Thermal Reorganization of a Protein Thermodynamics Structure

When a protein act stimulus, such as binding to ligand or temperature, the amplitude, scope, strength of the internal motion within a protein is changed, and this results in the reorganization of protein thermodynamic structure. In this process, the unique protein three–dimensional structure remains unchanged, but the conformation of a protein (or conformational state) is changed. The conformational change accompanied with enzyme-substrate binding has been well documented and has become common knowledge of biochemistry (8).

When enzymes act denaturants, the active conformation was first transformed into inactive conformation of enzyme and the enzyme activity was lost, but the global protein structure of enzyme remain unaltered. In this case, the transformation between active and inactive conformation of a protein is processed by thermodynamic reorganization of internal motions of a protein. If the concentration of denaturant is further increased, the protein will be denatured completely and irreversibly, the protein three dimensional structure will be destroyed. The general of protein denaturation and reorganization of protein thermodynamic structure is shown in Figure 3.

According to protein thermodynamic theory, the structural relation of diversified potherses of a protein is changeable with a changing condition, such as increasing of concentration of denaturant, or temperature. There is no systematic research about this topic of science, but we can judge its rightness by many fragmental evidences.

1) The competitive agonist could bind to a receptor, but it cannot induce the larger protein conformational change required for activation of the receptor (or reorganization of protein thermodynamic structure).
2) The potency of an agonist for a receptor is changeable. The interaction between agonist and environmental factors is needed for the maximum activity of the receptor. This indicated that the coupling efficiency between agonist binding conformations and active conformation of a receptor is changeable.
3) The protein thermodynamic structural reorganization is normally induced by protein interaction or protein complex formation. For example, the cooperation of hemoglobin is processed by interaction among different subunits.

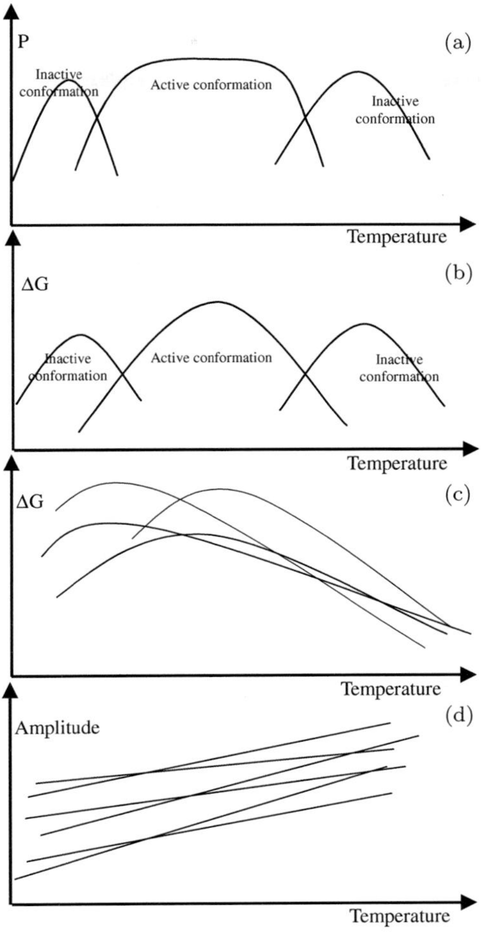

Figure 3. Diagram of Thermodynamic reorganization of protein.

Figure a is the distribution curve of global conformations of a protein. Figure b is the stability curve of global conformations of the protein. Figure c is the stability curve of local conformations (result from H-D exchange by Englander). Figure d is the amplitude of internal motion of atom which can be measured by NMR technique. As the change of internal motion within a protein, diversified conformations of a protein was appeared or generated under different conditions. In this Figure, we see that behaviors of protein conformation (or thermal motion) at diversified levels differs from each other greatly. The underlying reason is that information of internal motion of a protein has been integrated at diversified levels and the trait of protein conformation or dynamics at fundamental level may not be expressed in behavior of protein conformation at advanced level.

4) The activation energy of enzymatic reaction varies with species of substrate. It indicated the difference of coupling between substrate binding conformation and transition state conformation of enzymatic reaction.

References

1. Zhao, Q. (2001). Irreversible thermodynamics theory for protein folding and protein thermodynamics structure. *Prog. Biochem. Biophys.* 28, 429–435.
2. Zhao, Q. (2009). Protein thermodynamic structure. *IUBMB Life*, 61, 600–606.
3. Zhao, Q. (2011). Dynamic model for enzymes action, *Protein Pept. Lett.*, 18, 92–97.
4. Maity, H., Maity, M., Krishna, M. M., Mayne, L., & Englander, S. W. (2005). Protein folding: the stepwise assembly of foldon units. *Proceedings of the National Academy of Sciences of the United States of America*, 102(13), 4741–4746.
5. Krauss, M., Korr, D., Herrmann, A., & Hucho, F. (2000). Binding properties of agonists and antagonists to distinct allosteric states of the nicotinic acetylcholine receptor are incompatible with a concerted model. *J. Biol. Chem.* 275, 30196–30201.
6. Ricard, J., & Noat, G. (1985). Subunit coupling and kinetic cooperativity of polymeric enzymes. Amplification, attenuation and inversion effects. *J Theo Boil*, 117(4), 633–649.
7. Fields, P. A. (2001). Protein function at thermal extremes: balancing stability and flexibility. *Comparative Biochemistry and Physiology Part A: Molecular & Integrative Physiology*, 129(2), 417–431.
8. Isupov, M. N., Obmolova, G., Butterworth, S., Badet-Denisot, M. A., Badet, B., Polikarpov, I., . . ., & Teplyakov, A. (1996). Substrate binding is required for assembly of the active conformation of the catalytic site in Ntn amidotransferases: evidence from the 1.8 Å crystal structure of the glutaminase domain of glucosamine 6-phosphate synthase. *Structure*, 4(7), 801–810.

Chapter 13

General Properties of a Protein

Summary: The structural features and the biophysical natures of a protein are discussed on the basis of protein thermodynamic structure theory. We have demonstrated that much current knowledge about protein conformation and conformational change can be explained by applying the principles of protein thermodynamic structure theory.

Much knowledge of protein science has been obtained in the past, but there has not been a mature theory which can give reasonable explanation for it. This covers structural nature and physical properties of a protein and the logical relationship between them. The protein structure theory can give satisfied explanation to many problems of proteins science. Here in this chapter, we will not discuss these problems in detail. Rather, we will emphasize the general profile of it. Without knowing it, the real and comprehensive understanding of protein science cannot be obtained (1).

1. General Properties of Protein States

In view of physics, a substance is believed to exist in one of three discrete states: solid, liquid, and gas. There is no concept of fixed three-dimensional structure for liquid and gas, so a protein is generally considered at solid state because three dimensional structure theory of a protein could be measured experimentally. The concept of structural biology has been proposed and it states that protein function is fully determined by its three-dimensional structure.

However, some properties of a protein indicate that a protein is not in solid state (2–4). Unlike the solid state, which has fixed three dimensional structure, remains its shape and three dimensional structure when it acts a stimulus, such as pressure and temperature, when a protein can change its shape or conformation. It is a well-established belief that the protein conformational change has played an important role in protein functioning and regulation. Although a protein shows fixed three-dimensional structure in crystal state, a protein has no fixed three dimensional structures and its shape is changeable. Figure 1 shows two conformations of lysozyme which were obtained from different crystals.

In this Figure, we see that local protein conformation of lysozyme at one area (yellow or red) is near free and can adopt many configurations in solution. Or in other words, the protein flexibility at this area is much larger.

In crystal state of substances, all atoms of a protein are well fixed because in crystal state there is no free space which can permit the free motion of components of a protein. But in solution, the case is totally changed.

It is believed that a protein shows many different conformations in solution, but only one type of protein conformation is selected in the crystallization process of a protein.

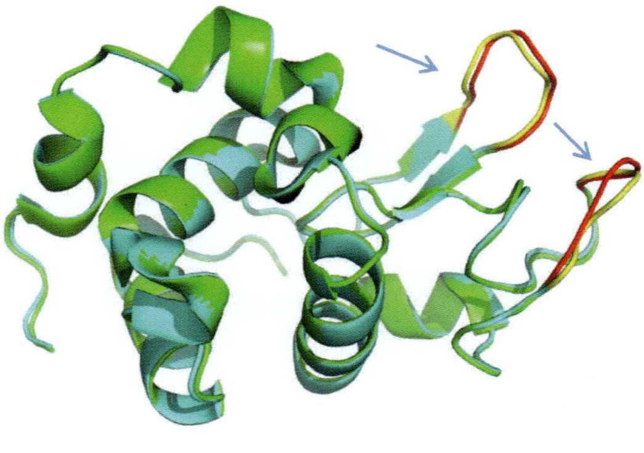

Figure 1. Superposition of two conformations of lysozyme.

The larger difference between the two protein conformations is marked as red and yellow. (This Figure was made by Dr. Yonglin Hu)

2. The Three-Dimensional Structure of a Protein is an External Imagery for Protein Thermodynamic Structure

A protein is a complex system of thermodynamics. The driving force to form a full fixed three-dimensional structure of a protein comes from the thermodynamic stability of it. Thus, the protein three-dimensional structure is an external imagery of protein thermodynamic structure theory.

If we assume that the minimum value of protein stability is Z, the decrease of protein stability would loosen the protein three-dimensional structure and the protein will show high flexibility.

In fact, we do not know the exact value of Z. But we know that all proteins have no fixed three-dimensional structure in solution even at its most stable condition.

The landscape is a good way to illustrate the relationship between protein stability and protein three-dimensional structure (5). The landscape of protein is shown in Figure 2.

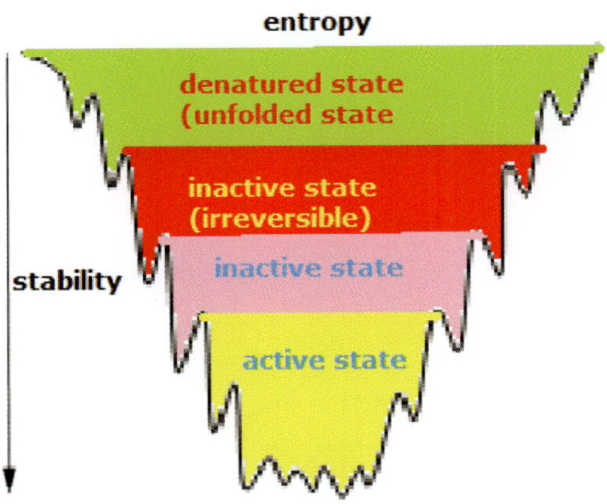

Figure 2. Landscape of a protein.

In this Figure, the protein flexibility could be described in some degree by entropy concept. High protein flexibility results in a large amount of entropy. The protein stability determines protein three-dimensional structure. The active state (or conformation) of a protein is stable and has been well structured. The inactive state has low stability and shows high flexibility. The molten state of a protein represents one special form of inactive conformation.

The protein flexibility or mobility of a protein has been well expressed in Figure 1. In this Figure, we see that protein stability determines the protein flexibility (or the structural nature of the protein). The structural freedom of a protein and its constituents increases when a protein stability decreases.

3. Structural Nature of a Protein

The structural freedom of a protein rises up through two different ways. One is within a potherse and the other is at interface of different potherses.

A simple system of thermodynamics comprises N particles, the number of (3N+1) relations (or interaction) is needed to give a full fixed three-dimensional structure.

Considering that a typical structure could be seen by some particles, not all particles, at least 12 (4×3) relations should be given. In a protein, 4 fixed residues within a protein can produce a dominating conformations within a potherse and other residues can be freely arranged in space. Let us remember that the maximum size of a potherse comprises 8 residues, in which two types of protein conformation is not justly permitted. The free residues within a potherse is a source of protein flexibility. Its principle is shown in Figure 3.

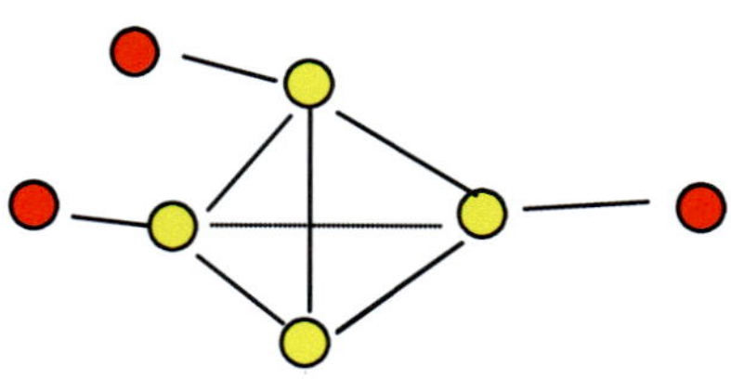

Figure 3. Diagram for the origin of the structural freedom of a protein.

Let us assume that a potherse (conformation) comprises 7 residues, 4 residues (yellow balls) forms a fixed three-dimensional structure and acts dominating structure, other residues (red balls) can freely be arranged in space. The conformational change occurs when old structure is destroyed and new structure is formed.

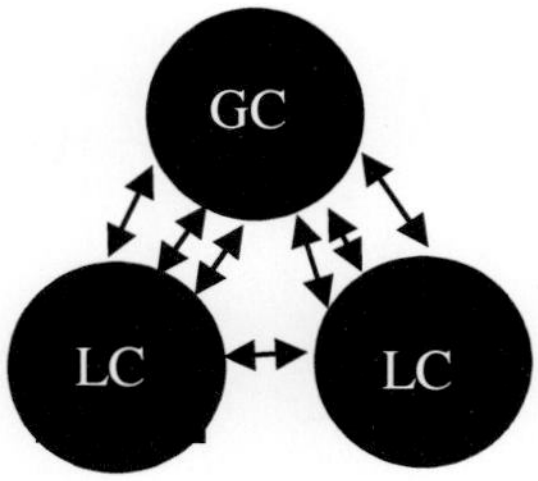

Figure 4. Diagram of structural features of a protein with hierarchical structure. GC represents global protein conformation, LC1 and LC2 represents two local conformations. To give a fixed structure between two conformations, like GC and LC1, three interactions between different conformations should be given. Three types of protein conformations are drawn in this Figure.

Unlike the crystal state of substance, a protein shows hierarchical structure, there is not any strict relationship between the advanced structure and the fundamental structure. This type of structural features can be clearly seen in following Figure.

The relations among different types of protein conformation is expressed in Figure 4. If local conformation is fixed up by its interaction to global conformation, the relationship between two local conformations cannot be fixed. This is another source of protein flexibility.

4. Diversified Changes Accompanied with Protein Conformational Change

In solution, the structural or conformational change of a protein cannot occur alone. The protein structural change or conformational change is accompanied with diversified change of solution and properties of a protein. These changes have been summarized in Figure 5.

In this Figure, we can see that protein conformation can link up all material, physical or chemical, mechanical or thermodynamic changes or properties in vivo. Proteins are processors of all types of bio-information and integrate them into cellular or molecular signaling activity.

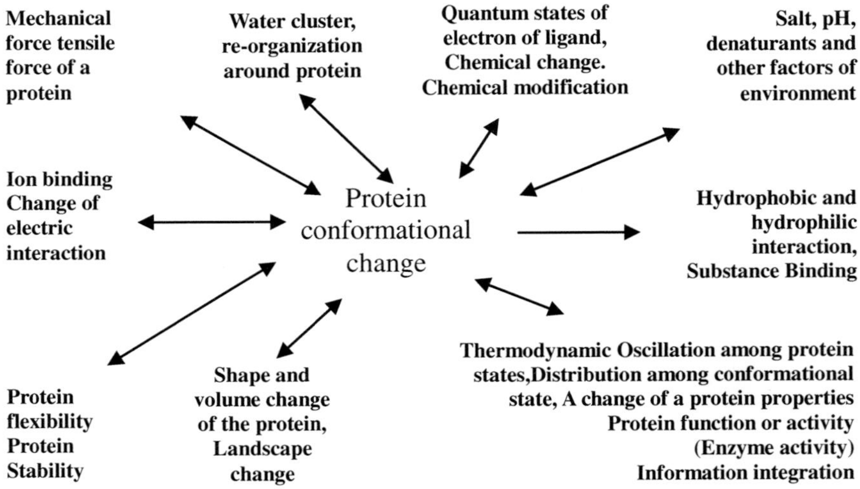

Figure 5. Proteins can link up all biological properties *in vivo.*

All possible relations among protein conformational change, physical and chemical properties of proteins is pictured in this Figure. For a protein, these relations may not be expressed under a given condition.

5. Multiple Dual Natures of a Protein

We have learnt that a protein has no fully fixed three dimensional structure in solution and there is structural freedom within a protein. Theoretically speaking, protein three-dimensional structure determines their nature of solidity and a change of such structures can result in a change into liquid. Therefore, the state of a protein stays somewhere between solid and liquid. In protein science, this nature is studied under the name of protein flexibility or protein dynamics.

We have learnt that the partition function of the protein conformational state is environmentally dependent,

$$N_{ps} = \exp\left(-\frac{[\Delta G^0 - \mathrm{cs}]}{kT}\right).$$

Thus, we get a conclusion that a protein shows multiple natures. We summarize it as follows.

Dual natures of solid and liquid material: The solid nature of a protein can be shown in the following aspects: 1) it has three-dimensional structure; 2) its behaviors can be described by laws of mechanics. The liquid nature of a protein is fully shown in the protein conformational change.

Dual properties of mechanics and thermodynamics: a protein can respond mechanical and thermodynamic stimulus. The mechanical force, voltage, pressure can change conformation of a protein. A change of internal thermal motion of protein can also alter protein conformational state. One phenomenon of protein, which has not been reported, is the separation of protein subunits induced by electric field. It has been observed that weakly bound protein complex can be separated by very low electric field intensity, say, 2 v/cm, the antibody-antigen complex can be separated at high intensity of electric field, say 70 v/cm. The DNA stands, one stand of which is fixed, can be separated at rather low intensity of electric field and low temperature. These tests confirm the concept that protein structure is sensitive to mechanical force. These concepts make us analyze properties of a protein by applying laws of mechanics and thermodynamics under different cases (1).

Dual properties of macroscopic and microscopic substance: a protein can act as one particle in solution and this is a property of microscopic substance. A protein comprises many constituents and its shape can respond to mechanical force and this is a property of macroscopic substance.

Dual properties of hydrophilic and hydrophobic interaction within a protein: it is well established conclusion that protein stability comes from hydrophobic interaction among constituents of a protein, but the protein three-dimensional structures are formed and maintained by hydrophilic interaction or H-bond. In the molten state of a protein, the protein three-dimensional structure is lost, but it still shows thermal stability.

Dual natures of specific interactions and unspecific interactions within a protein: theoretically speaking, hydrophobic interaction among different substances is an unspecific interaction and it can be replaced with substance with similar properties, the hydrophilic interaction belongs to specific interaction among different substances and the structure or conformation of the substances has a great impact on the specific interaction. When we consider the interaction between protein and other substances, we can find that protein denaturants, such as urea, ethanol, can change protein conformation by unspecific way; in contrast, the ligand of a protein can bind to specific site of a protein and it changes protein conformation by a specific way.

Overall, the multiple natures of a protein endow the complex properties of a protein and this make protein a versatile processor for bio-information.

References

1. Zhao, Q. (2013). Nature of protein dynamics and thermodynamics. *Reviews in Theoretical Science*, 1(1), 83–101.
2. Zhu, L., Sage, J. T., Rigos, A. A., Morikis, D., & Champion, P. M. (1992). Conformational interconversion in protein crystals. *Journal of Molecular Biology*, 224(1), 207–215.
3. Fenimore, P. W., Frauenfelder, H., McMahon, B. H., & Parak, F. G. (2002). Slaving: solvent fluctuations dominate protein dynamics and functions. *Proceedings of the National Academy of Sciences*, 99(25), 16047–16051.
4. Parak, F. G. (2003). Physical aspects of protein dynamics. *Reports on Progress in Physics*, 66(2), 103.
5. Thielges, M. C., Zimmermann, J., Yu, W., Oda, M., & Romesberg, F. E. (2008). Exploring the energy landscape of antibody — antigen complexes: protein dynamics, flexibility, and molecular recognition. *Biochemistry*, 47(27), 7237–7247.

Chapter 14

Protein Conformation

Summary: In the field of biochemistry and molecular biology, much knowledge of protein science has been explained on the basis of the concept of protein conformation. Here, we will study concept of protein conformation on the basis of protein thermodynamic structure theory. The analysis shows that a protein can show numerous types of protein conformations under different conditions.

In the field of biochemistry and molecular biology, our understanding of protein science was principally established on two concepts: protein conformation and protein flexibility. Theoretically and originally speaking, protein conformation represents the spatial configure of constituents of a protein and it is another name for protein three-dimensional structure. Without any scientific definition, most scientists use protein conformation concept to describe dynamic configure of atoms of a protein and protein three-dimensional structure refers to static configure of residues of a protein. So, protein conformation may represent a transient configure of atoms in space and it may have no fixed three-dimensional structure. In contrast, the uniqueness protein three-dimensional structure refers to the fully fixed and the permanent configuration of a protein.

Although in many textbooks of biochemistry, the view that protein function is determined by its three-dimensional structure is emphasized, the protein function is logically related protein conformations and its change (1, 2). In the field of biochemistry, protein conformation also refers to protein conformational state or thermal state of a protein. In this chapter, we will discuss protein conformation concept in two aspects: static and dynamic (or conformational change).

1. Protein Conformation Concept

In the 1950s, the protein conformation concept has been introduced into the field of biochemistry for scientists who had known that the shape and thermal states of a protein is changeable; at that time, the protein conformation refers to protein three-dimensional structure (3, 4). After the protein three-dimensional structure theory has been fully established in 1960s, we learned that the unique protein three-dimensional structure is unchanged for native states of a protein. As protein unique structure theory has gradually dominated our understanding of protein science, our understanding of protein conformation concept has changed. In order to differentiate the concepts of protein conformation and the unique protein three-dimensional structure, the meaning of protein conformation has been narrowed and now it refers to temporal configures of constituents of a protein. It is said that a protein has its unique three-dimensional structure which remains unchanged unless the protein is denatured, but its conformation is changeable in solution. Gradually, concepts of global and local conformation, inactive and active conformation of enzyme, are introduced into the field of protein science, the studying of protein conformation and its change becomes an important task of biochemistry. Many valuable views of biochemistry come from such studying, such as the activation mechanism of receptor and the dynamic model for enzyme action.

Because many experimental and theoretical research methods have been applied into protein science, our understanding of protein conformation concept differs largely and principally.

For most biochemists, the protein conformation concept refers to the protein conformational state or the thermal state of a protein. The actual meaning of it can be judged by the properties by which it is measured. Scientists have studied the behaviors of enzyme activity, receptor activity, and protein functions, the functional change of a protein can be observed in some cases and it is hypothesized that such a change is induced by protein conformational change.

In this case, it refers to protein conformational state although protein conformational change can be measured experimentally. The key difference between protein conformational state and protein conformation is that the energetic level of a protein conformational state is changeable.

A different view arises up from the study of protein dynamics. There are numerous internal motions for a protein; the position of atoms within a protein varies by time, and the configure of these atoms in space and time can be only described by concept of ensemble of protein conformation (5–7). In this case, the protein conformation represents subtle and transient local configure of atoms of a protein. Often these types of conformation are far out of the scope of protein function, biochemistry and molecular biology. But this study on this topic has revealed the dynamic nature of protein conformational change.

2. Experimental Measurement of Protein Conformation

In order to monitor and detect protein conformational change, scientists in the field of physics and chemistry have explored a number of experimental methods. One experimental method can only provide one type of information of protein conformation or its change, and the overall view about protein conformation should be obtained by analyzing and integrating all types of information we have obtained.

The static protein conformation can be measured by x-ray diffusion method. When a protein is crystallized under different conditions, different conformations of a protein may be obtained. It takes an advantage that it can provide undoubted evidence for protein conformation, but most types of protein conformational change cannot be measured by this technique.

The CD (circular dichroism) and other methods of spectrographic analysis, such as fluorescence spectrum, have been widely utilized in biochemistry (8, 9). The CD technique has become the standard method for judging global protein conformational change related to protein stability. The main flaw of such method is that it is not sensitive to local conformational change of a protein.

The protein conformation stability is a measurement for protein conformational change. The H-D exchange test has provided valuable information for local protein conformation (10).

The most sensitive and versatile methods to judge protein conformational change is the measurement of a change of receptor or enzyme activity. Although this method cannot give any direct information of protein conformation or its change, it can be utilized to judge the protein conformational change for any change of enzyme activity resulting from a change of protein conformational change. In fact, much valuable about receptor conformation comes from such studying (11, 12). In these cases, the protein conformation refers to the conformational state or the thermal state of a protein. For example, the agonist binds to a receptor and the agonist-binding conformations of a receptor is isolated. There are differences between the agonist binding affinity and the potency of agonist. And then the active conformation of receptor for agonists binding can be further isolated.

There are many other methods which cannot be discussed here. But there is no method which can give all the information about protein conformation and its change.

3. A Protein Shows Infinite Types of Protein Conformation

When protein conformation concept has been proposed, we should ask how many types of protein conformation exist within a protein and how to count it.

By applying system logic theory, we could get one conclusion: the protein conformation could be generated even for a folded protein (13). When a protein acts a stimulus, a new type of conformation could be generated. See Figure 1 for illustration.

For a given condition, there is a few type of protein conformation at global level. There are numerous types of conformation at the atom and the local level. In the field of protein dynamics and computational biology, the ensemble of protein conformation concept has been proposed to describe this property.

4. The Size or Spatial Scale of Protein Conformation

According to protein thermodynamic structure theory, a potherse is composed of motions of many residues; the upper limit is 16 independent motions, which corresponds to 8 free residues. In a folded protein, the motion of amino acid residue is limited and the number of actual residues within a pother may vary from 4-32 (13).

The foldon concept, proposed by SW Englander, is another expression of a potherse. According to SW Englander, a foldon represents a cooperative unit of thermodynamics in protein conformational change. The data obtained from his lab agrees with our prediction well (14).

A highlighted feature of these residues is that they may be located in different sites within a protein. This trait can be easily understood by the concept of potherse or foldon. In this case, the protein conformation represents an abstract concept and it does not correspond to a protein three dimensional structure in any way.

As we can see in Figure 1 of this chapter, A and B are constituents of protein conformation for ligand binding, they are at different position in protein sequence as well as in space. In this case, the local protein conformation refers to a state that structural change at other sites is unnecessary. If structural change at other sites is needed to form a new conformation, it is at global level.

Figure 1. The generation of new protein conformation.

In this figure, A and B represent two sites of a protein. When this protein has not bound to a ligand, the motion of A and B are independent. When the protein binds to its ligand, the motion of A and B are coupled via the ligand. In other words, a new type of protein conformation has been generated.

5. Potherse Change as Molecular Switch

According to protein thermodynamic structure, there are two types of protein conformational change; change of pother and change of potherse. The pother change represents fundamental and continuous change of internal motion within a protein, the labeling parameters of a pother changes continuously, such as amplitude of thermal motion of atoms within a protein. The change of potherse represents structural change of a protein, both in thermodynamic structure and three-dimensional structure; or in other words, change of potherse represents a change of the thermal state of a protein. The change of potherse is processed catastrophically. When change of potherse occurs, a dramatic change of protein properties (or discontinuous change of protein properties) could be observed or measured experimentally.

The protein denaturation is a specific process of potherse change. The receptor activation is another example of potherse change. The principle of potherse and pother change is shown in Figure 2.

The protein conformational change in denaturation is shown in Figure 2. When the concentration of urea is increased, the

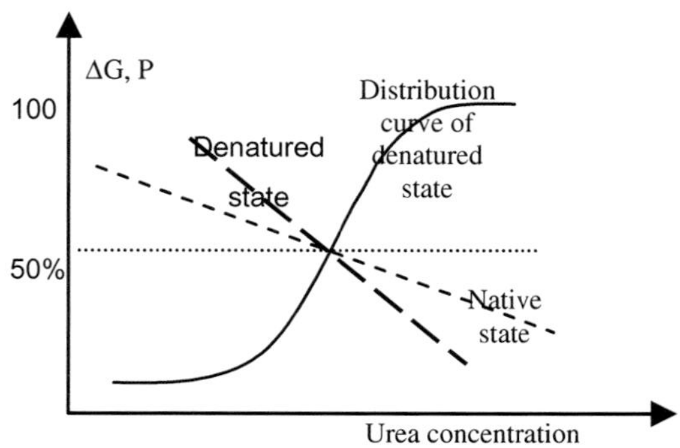

Figure 2. Potherse change of a protein.

ΔG is stability and P is probability of protein conformation. In this figure, the stability curve of denatured state (potherse) and native state of a protein (another potherse) is shown. When the energy difference between two states reaches zero, the probability of two states is 50%. In views of equilibrium thermodynamics, the protein can be either at native or denatured state and there are no intermediate states between two states.

energy level and the amplitude of internal motion of a protein also increased, and the stability of a protein is decreased. This change of protein conformational change belongs to pother change. When concentration of urea is further increased, the protein begins denaturation. If the protein conformational change is reversal, the probability of denatured state or native state can be described by thermodynamic laws. The transformation between native and denatured state of a protein belongs to the potherse change.

As potherse change represent thermal state change, it acts as a molecular switch in any biochemical or molecular reaction (13).

6. Complex Molecular Switch

Another switch model that has been widely accepted by experimental biochemists is allosteric cooperativity (15). The sigmoid binding curve of oxygen to hemoglobin was observed by Bohr in 1904 and he proposed that the binding of first oxygen molecule to hemoglobin is helpful for the binding among other oxygen molecule and receptor, or in other words, the binding of 4 oxygen molecules to hemoglobin is cooperatively. The allosteric mechanism of enzyme regulation was soon established on the basis of studying on this topic. Behaviors of positive and negative cooperativity were shown in Figure 3.

For enzyme with positive cooperativity, the binding of first molecule of substrate to enzyme is helpful to the binding of next molecule of substrate to enzyme. For enzyme with negative cooperativity, the binding of first molecule of substrate made it difficult to bind the next molecule of substrate. For enzyme with cooperativity, the maximal sensibility (dv/dc) is at one concentration of substrate. For enzyme with positive cooperativity, the precise enzyme regulation with the maximal sensibility (dv/dc) occurs in a range of substrate concentration around the k_m of enzymatic reaction. For enzyme with negative cooperativity, the precise enzyme regulation occurs in a range of substrate concentration below the k_m.

This type of molecular switch is very complex and it is not the basic definition of molecular switch. In fact, it comprises two molecular switches at different levels, the binding of enzyme and

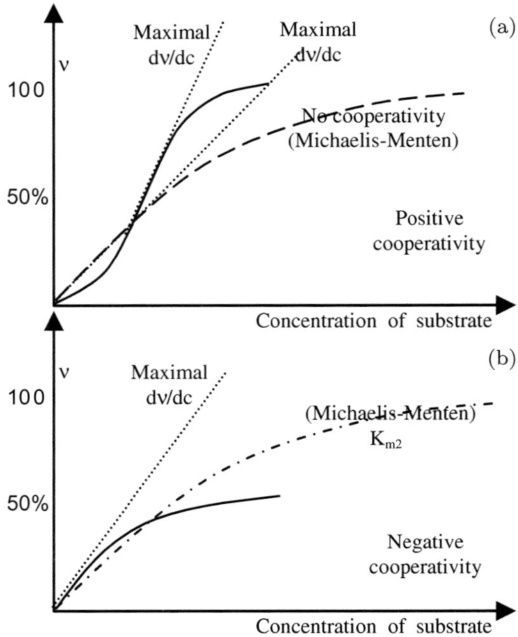

Figure 3. Negative and positive cooperativity expressed in enzymatic kinetics. The kinetic behavior of enzymatic reaction with positive cooperation is shown in Figure 3A. Figure 3B is the profile of negative c cooperativity.

substrate act as a molecular switch at fundamental level and which determines the order of magnitude of substrate concentration of enzyme regulation; the interactions among different subunits of enzyme subunit acts as a molecular switch at an advanced level which determines the precise range of substrate concentration for regulation and nature of cooperativity.

7. Stability Curve of Protein Conformation

The temperature is a fundamental biophysical factor that influences protein conformations and functions. Thus, studying the conformation distribution along the temperature gradient is of great importance. First, Cp, which represents the specific heat of one conformational state, is assumed to remain unchanged in all temperature ranges. The specific heat difference between two states is ΔCp. If a suitable reference state is selected ΔCp can be positive.

Changes in the Gibbs free energy, enthalpy, and entropy (ΔG, ΔH, and ΔS) between the two states of a protein at varied temperature can be calculated as follows:

$$\Delta H = \Delta H^0 + \Delta C_p(T - T^0)$$

$$\Delta S = \Delta S^0 + \int_{T^0}^{T} \left(\frac{\Delta Cp}{T}\right) dt = \Delta S^0 + \Delta C_p \ln\left(\frac{T}{T^0}\right)$$

$$\Delta G = \Delta H^0 - T\Delta S^0 - \Delta C_p(T - T^0 - T\ln(T/T^0)$$

Where T^0 is the reference temperature, ΔH^0 is the enthalpy difference between the two states at T^0, ΔS^0 is the entropy difference at T^0, and ΔG^0 is the Gibbs free energy difference between the two states at T^0.

Middle Equation is called the Gibbs-Helmholtz equation. In protein science, Becktel and Schellman firstly deduced the equation by applying it to protein denaturation (16). This equation represents the protein conformation stability curve. From the partition function of protein conformational state and protein stability curve, the conformation distribution curve along the temperature gradient can also be described by this equation.

The general profile of protein conformation stability curve is described in Figure 4.

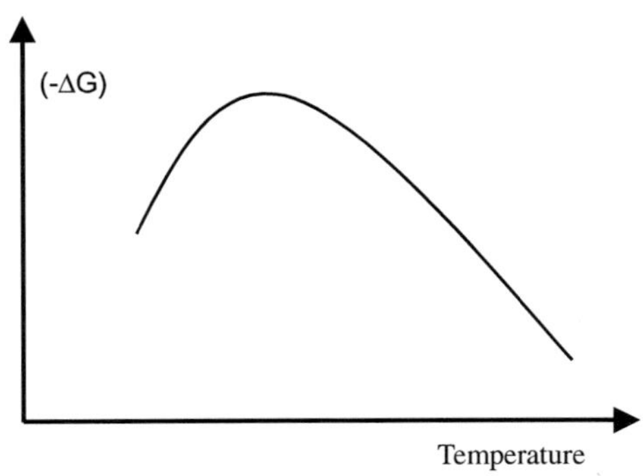

Figure 4. Protein conformation stability curve.

Englander *et al.* confirmed that the stability of a local conformational state of a residue within the protein can be adequately described by the protein conformation stability curve (17).

8. The Thermodynamic and Dynamic Stability of Protein Conformation

The protein thermodynamic stability has been widely studied experimentally. It has shown that the protein stability is rather low; most are in a range of 5–15 kcal/mol. In view of chemistry and thermodynamics, this energy barrier is so little and a protein would not be stable in solution.

Although the thermodynamic stability of a protein is rather low, the dynamic stability is very high. It has shown that the activation energy for protein denaturation is very high and it can be more than 100 kcal/mol. So, the stability of a protein comes from its dynamic stability.

The protein conformation stability of thermodynamics has not been studied experimentally, but it must be lower than that of protein stability. But we can judge the range of it according to distribution curve of active conformation of receptor or enzymes. The following profiles of distribution curve could be easily observed in numerous scientific report and we will analyze it theoretically.

When the difference of stability in active and active conformation reaches zero, the probability of active conformation of a protein is 50% if it distributed between two states; it would be slightly lower than 50% if it distributed among many states. Therefore, the conformation stability of a protein expressed in Figure 5A is rather small, and its value is about zero. The conformational stability can be measured experimentally, but more data is needed.

Finally, we have summarized the energy level of protein conformational change (thermal and dynamic stability) in Figure 6.

9. The Timescale of Protein Conformational Change

The energy involved in protein conformational change is normally rather small, and some properties of protein conformational change

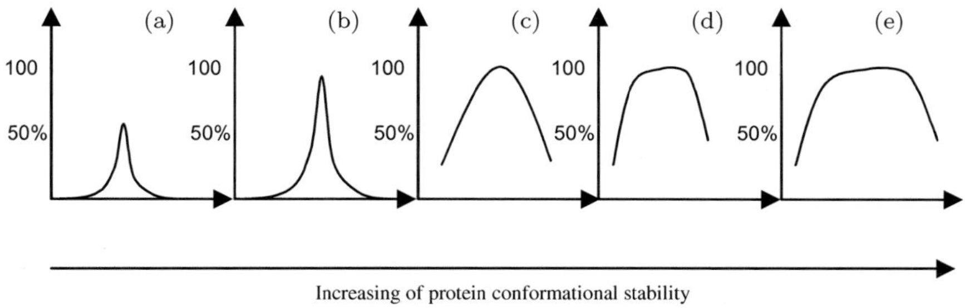

Figure 5. Profiles of protein conformational distribution curve.

As the increase of protein conformation stability (ΔΔG, difference between stabilities of active conformation and inactive conformation of a protein), from A to E, the profile of distribution curve of protein conformation is changed.

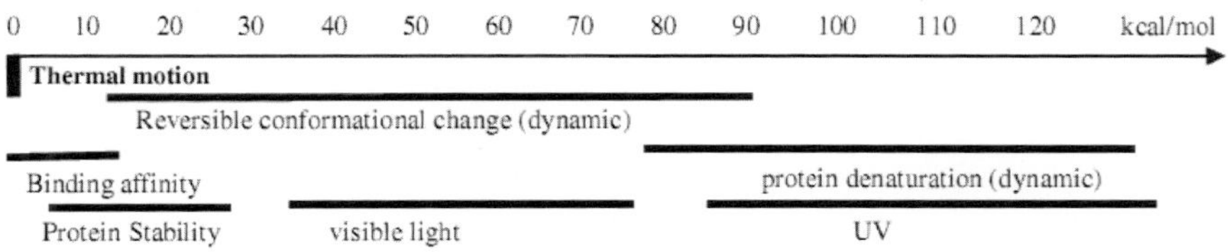

Figure 6. The energetic level of protein conformational change.

cannot be analyzed by the concept of Gibbs free energy. The timescale of protein conformational change is an important measurement for it and many properties of protein conformational change can be analyzed by this concept (18).

A highlighted nature of protein conformational change is that the rate of protein conformational change is not determined by Gibbs free energy. In common view of traditional chemistry, the rate of chemical reaction will be increased if the Gibbs free energy of the chemical reaction (or chemical potential) becomes stronger. But in protein conformational change, the rate of protein conformational change remains unchanged and the timescale will be increased if the Gibbs free energy of protein conformational change is increased (its mechanism will be discussed in Chapter 16).

The timescale of protein conformational change is logically related to protein function and properties; we have summarized some types of protein conformational change in Figure 7.

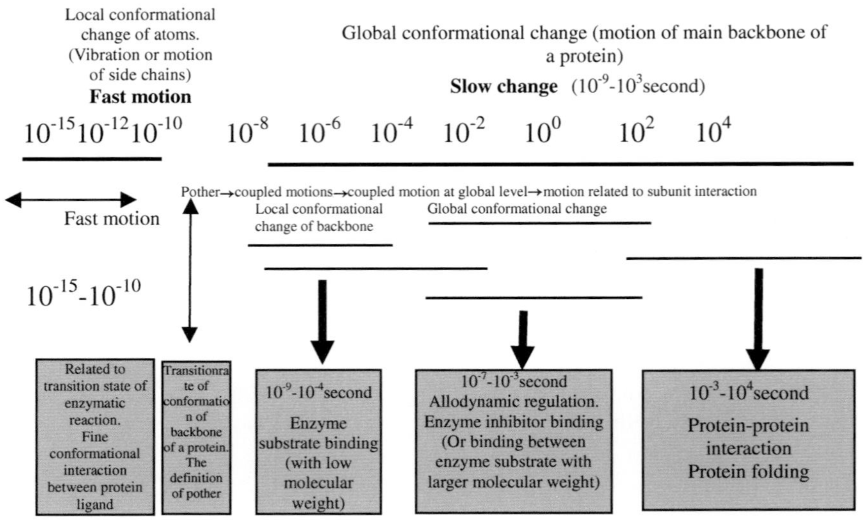

Figure 7. The timescales of protein conformational change and related functions.

References

1. West, E. S., Todd, W. R., Mascon, H. S., & Van Bruggen, J. T. (1974). *Textbook of Biochemistry*. Oxford and IBH Publishing.
2. White, A., Handler, P., Smith, E., & Stetten Jr, D. (1959). Principles of biochemistry. *Principles of Biochemistry.*, (Edn 2).
3. Lumry, R., & Eyring, H. (1954). Conformation changes of proteins. *The Journal of Physical Chemistry*, 58(2), 110–120.
4. Koshland Jr, D. E. (1958). Application of a theory of enzyme specificity to protein synthesis. *Proceedings of the National Academy of Sciences of the United States of America*, 44(2), 98.
5. Frauenfelder, H., Sligar, S. G., & Wolynes, P. G. (1991). The energy landscapes and motions of proteins. *Science*, 254(5038), 1598–1603.
6. Shaw, D. E., Maragakis, P., Lindorff-Larsen, K., Piana, S., Dror, R. O., Eastwood, M. P., ... & Wriggers, W. (2010). Atomic-level characterization of the structural dynamics of proteins. *Science*, 330(6002), 341–346.
7. Boehr, D. D., Nussinov, R., & Wright, P. E. (2009). The role of dynamic conformational ensembles in biomolecular recognition. *Nature Chemical Biology*, 5(11), 789–796.
8. Greenfield, N. J., & Fasman, G. D. (1969). Computed circular dichroism spectra for the evaluation of protein conformation. *Biochemistry*, 8(10), 4108–4116.

9. Royer, C. A. (2006). Probing protein folding and conformational transitions with fluorescence. *Chemical Reviews*, 106(5), 1769–1784.
10. Huyghues-Despointes, B. M., Pace, C. N., Englander, S. W., & Scholtz, J. M. (2001). Measuring the conformational stability of a protein by hydrogen exchange. In *Protein Structure, Stability, and Folding* (pp. 69–92). Humana Press.
11. Kenakin, T. (2001). Inverse, protean, and ligand-selective agonism: matters of receptor conformation. *The FASEB Journal*, 15(3), 598–611.
12. Totrov, M., & Abagyan, R. (2008). Flexible ligand docking to multiple receptor conformations: a practical alternative. *Current Opinion in Structural Biology*, 18(2), 178–184.
13. Zhao, Q. (2009). Protein thermodynamic structure. *IUBMB Life*, 61(6), 600–606.
14. Maity, H., Maity, M., Krishna, M. M., Mayne, L., & Englander, S. W. (2005). Protein folding: the stepwise assembly of foldon units. *Proceedings of the National Academy of Sciences of the United States of America*, 102(13), 4741–4746.
15. Koshland, D. E., & Hamadani, K. (2002). Proteomics and models for enzyme cooperativity. *Journal of Biological Chemistry*, 277(49), 46841–46844.
16. Becktel, W. J., & Schellman, J. A. (1987). Protein stability curves. *Biopolymers*, 26(11), 1859–1877.
17. Bai, Y., Milne, J. S., Mayne, L., & Englander, S. W. (1994). Protein stability parameters measured by hydrogen exchange. *Proteins: Structure, Function, and Genetics*, 20(1), 4–14.
18. Zhao, Q. (2013). Nature of protein dynamics and thermodynamics. *Reviews in Theoretical Science*, 1(1), 83–101.

Chapter 15

Dynamic and Thermodynamic Nature of Protein Conformational Change

Summary: Each protein is a complex system of thermodynamics and its behaviours differ from that of common materials in principle. If we use our knowledge of conventional physics when studying protein science, there may be mistakes and controversies. In fact, some thermodynamic natures of protein conformational change have been classified into the dynamic nature of it. Our analysis shows that the dispute about the role of protein dynamics in enzyme activity and protein regulation results from the wrong understanding of dynamic and thermodynamic nature of protein conformational change. There is a direct relation between protein thermodynamics and enzyme activity (a type of protein functioning); the protein dynamics takes its role in enzyme activity by influencing thermodynamic properties of protein conformational states.

A protein is a complex thermal system and protein conformational changes are logically composed of many internal motions and vibrations of atoms and amino acid residues (1–5). In order to explain protein properties theoretically, one should know the dynamic and thermodynamic natures of protein conformational change. If the thermodynamic nature of a protein is mistakenly attributed to the dynamic nature of a protein, and vice versa, wrong conclusions may be produced.

Protein dynamics is a rather confusing term in science. To simply put, it represents the internal motion or vibration of components of a protein. It is not the synonym of the dynamic nature of a protein. However, it can both refer to dynamic or thermodynamic nature of a protein under different cases, hence we must study it carefully.

Enzymes represent a special class of protein and it has been documented that protein dynamics are related to the execution of enzyme activity (6–9), but their underlying mechanism is not clear in the past and there is still heated debate about it (10–12). Currently, there are three different views about it. The first view is that protein dynamics (or internal motion) contribute much to the high efficiency of enzyme activity. Another view is that protein dynamics have nothing to do with enzyme activity and that the principles of thermodynamics can account for all natures of enzymatic kinetics (13–16).

The third view, or our theory about dynamic model of enzyme action, has been developed on the basis of protein thermodynamics, and it provides a new way to understand the role of protein dynamics in enzyme activity and protein functioning. It states that protein dynamics is essential for the execution of enzyme activity, but it takes its role by altering properties of protein thermodynamics.

Here we have shown that the misunderstanding of the dynamic and thermodynamic nature of a protein results in the debate about the role of protein dynamics in the execution of enzyme activity. Thus, it is necessary to carefully study the dynamic and thermodynamic nature of a protein, and to recheck our thoughts about it, along with enzymatic kinetics.

1. Scientific Criteria for Protein Dynamics and Thermodynamics

Theoretically speaking, the criteria for dynamic and thermodynamic nature of a protein is clear.

If $\mathrm{d}p/\mathrm{d}t = 0$

where p represents macro-world properties of a thermal system such as pressure and force; or

$P_{ro} = \text{constant}$

where P_{ro} represents the probability of a thermal state;

then the target property is the thermodynamic nature of a system.

If dp/dt is not zero,

then p represents the dynamic nature of a system.

These criteria are theoretical, absolute, and final for judging any disputes.

However, such criteria are too theoretical and most scientists do not adopt them in their studies. They prefer to adopt experimental criteria which appropriately match up with their studies for judging.

Experimental scientists point out that if a physical property cannot be described by the law of equilibrium thermodynamics, it can be recognized as the dynamic nature of a thermal system and vice versa. However, mistakes can be made when the thermal laws are invalidated under certain environments.

2. Differences between Simple and Complex Systems of Thermodynamics

According to the equipartition theorem of thermodynamics, for simple thermal system which has been well documented in textbook of physics, there are equations:

$$\mathrm{dE/dt} = \text{constant}$$

$$\mathrm{dE/dS} = 0$$

where E is average energy of a part of the thermal system, where S represents the quantity of environmental factors, such as pressure or viscosity, (but not temperature), and where t is time.

But for a complex system such as a protein, it is another matter. According to partition function of protein conformational state,

$$N_{ps} = \exp\left(-\frac{[\Delta \mathrm{G}]}{kT}\right) = \exp\left(-\frac{[\Delta G^0 - \mathrm{cs}]}{kT}\right)$$

then

d(ΔG)/dt is not the constant

d(ΔG)/dS is not zero.

So, if we utilize the criteria of simple thermal system in the field of protein science, mistakes will be made.

A highlighted property of a protein is that the internal energy of its components is a function of environmental factors. In contrast, the internal energy of components of simple thermal system is a function of temperature only.

In fact, as we have discussed, many mistakes have been generated and the debate about the role of protein dynamic in enzyme activity has emerged from such mistakes.

3. Experimental Concepts Related to Dynamic and Thermodynamic Nature of a Protein

When the criteria of dynamic and thermodynamic nature of a protein have been established, the next step is to recheck the concepts of protein science and ascertain the nature of these concepts.

Many different concepts describing the mobility of a protein have been utilized in diverse scientific fields. These concepts include protein flexibility, protein dynamics, internal motion, pother, and potherse (13).

Protein dynamics is a mechanistic concept mainly utilized by biophysicists who study thermal motions or vibrations within a protein (1, 4, 17). Theoretically, it refers to any internal motion or vibration of a protein. The internal motion of a protein is adopted by scientists who study protein dynamics by applying the NMR method which can directly measure the amplitude and range of internal motion of a protein (18). The protein conformational change, which represents coupled motions within a protein, at diversified levels or ranges of a protein cannot be measured experimentally and analyzed by applying the aforementioned concepts. Thus, biochemists prefer to utilize the term protein flexibility.

As a mechanistic term, the protein dynamics and internal motion of a protein do not always represent the dynamic nature of a protein. The thermodynamic nature of a protein refers to its statistics behaviors and dynamic nature, which represent its mechanistic behaviors. If one claims that protein dynamics contribute to enzyme activity directly, we should demonstrate that enzyme activity is a mechanistic process, rather than a thermal process or statistic behavior of a thermal process. As enzymatic kinetics can be precisely described by thermal model or Michaelis–Meten equation, the dynamic model for enzyme action can be ruled out completely.

Protein flexibility is widely utilized in biochemistry and molecular biology (5, 19–23). This term originally refers to the structural deviation of a protein (or conformation) from its standard structure (or unique structure). Thus, it refers to structural flexibility. However, it has been gradually utilized in the thermodynamic or mechanical deviation of a protein from its standard state (22, 23). This concept can represent both the dynamic and thermodynamic nature of a protein under different cases. But in the field of biochemistry, the protein property is normally measured at equilibrium state, thus most cases represents the thermodynamic nature of a protein.

Pother and potherse are theoretical and abstract concepts that are not concerted with experimental research methods. The logical properties of coupled motion can be theoretically analyzed by these concepts; a logical deduction system for protein thermodynamics has thus been developed (2, 13). It also refers to the thermodynamic nature of a protein.

4. Protein Dynamics is Related to Enzyme Activity Indirectly

After the protein flexibility concept was proposed by Koshland, the relation between enzyme action and protein dynamics has attracted much attention and this topic has been verified by many methods (1–5, 9, 21, 25). There is no consensus, but serious debates about it. In order to resolve such debate, the re-evaluation of scientific evidence of it is necessary.

Enzyme activity is influenced when urea (or other ions and solvent) was added in the working buffer of enzymes. The change of the enzyme activity was previously believed to be induced by a change of protein flexibility (9). However, a careful analysis reveals that such change in enzyme activity is induced by a change in the abundance of active conformation of the working buffer enzyme, so it represents the thermodynamic nature of enzyme action (9, 19, 26). The change in protein dynamics is a prerequisite condition in these cases, but it is not sufficient for a change in enzyme activity (16). Similar mistakes occur in studying the temperature adaptation of enzymes (24, 27). Thus, this is not the most direct and powerful evidence for the debate.

The measurement of kinetic isotope effect of enzymatic reaction is a sensitive method to diagnose the fine conformational state involved in enzymatic reaction. Many scientists have explored this technique before. Klinman found that the transition state of enzymatic reaction is a dynamic structure which is related to the protein conformational change in enzymes (28, 29). It was used as an evidence for the logical relation between protein dynamics and enzyme action. However, she did not consider the thermodynamic mechanism in her study. My theory states that the energy level of active conformational state of enzymes and the distances between different atoms within a protein are changeable under different temperatures due to the change of the protein thermodynamic state. It is the thermodynamic nature of protein conformational state involved in enzymatic reaction, rather than its dynamic nature.

Kern *et al.* have utilized NMR technique in their study of the abovementioned property (17). However, the coupled motion and protein conformational changes were not addressed in their paper, so their results are merely indirect evidence. The co-change in protein dynamics and enzyme activity could be argued as superficies. If there is any direct relation between internal motion of enzymes and enzyme activity, one should find out its meanings in concrete parameters of enzymatic kinetics and quantitatively express these parameters in the Michaelis–Menten equation. However, the aforementioned researchers failed to do so.

The promoting motion or vibration concept proposed is a valuable idea (7). This concept states that certain types of internal motion or vibrations of atoms (or residues) of enzymes play a crucial role in the catalyzed process of enzymatic reaction. In my opinion, motions of dynamic sites of enzyme action regulate the protein conformation or thermodynamic state of the transition state of enzymatic reaction; by this way, they indirectly accelerate the rate of enzymatic reaction. Misled by the traditional transition state theory of enzymatic reaction, some scientists have attempted to find direct relation between promoting motion and enzyme activity, which is questionable. Therefore, these results are merely indirect evidences for the relation between protein dynamics and enzyme activity.

5. Protein Dynamics Plays Its Role in Enzyme Activity by Influencing Protein Thermodynamics of Enzymes

In order to demonstrate the logical relation between protein dynamics and enzyme activity experimentally, one should apply following methods: (1) Given that enzyme activity is a non-concrete physical property, we should provide a physical definition of enzyme activity on the basis of a concrete physical property of protein dynamics; (2) measure the protein dynamics (or flexibility) at diverse levels and determine its quantitative relation with enzyme activity.

The scientific criteria, research method, and logical relations between different concepts of science can be only demonstrated by a new theory, and experimental science is unable to do so. Thus, we cannot confirm such a relation experimentally without a new theory in which the logical relation between protein dynamics and enzyme activity can be demonstrated logically.

After obtaining such a conclusion in 1987, we completely ceased all efforts to experimentally confirm this issue and tried to explain it theoretically. We proposed a basic hypothesis in 1996, which states that any protein function or enzyme activity correspond to a potherse within a protein (6, 15). Thus, the logical relationship between protein dynamics and enzyme activity was established theoretically. But a potherse represents a thermal system which is ladled by a parameter of protein thermodynamics, and thus within our theory, there is an indirect relationship between protein dynamics and enzyme activity.

6. Revised Michaelis–Menten Model

The theoretical foundation of enzymatic reaction is the Michaelis–Meten equation proposed by Michaelis and Menten in 1913 (30). If there is a relation between protein dynamics and enzyme activity, we should find the essential relationships between parameters of enzymatic reaction and protein dynamics.

The interpretation of the Michaelis–Meten equation by the transition state theory is a remarkable process in enzymology. Pauling proposed an important conclusion and inferred that the enzyme binds to the transitional state of the reactant more tightly (31). The transition state theory for enzymes is clearly correct in certain aspects, but the role of protein conformational change involved in enzymatic reaction cannot be analyzed by this theory. It is well known that rate-limiting step in some enzymatic reactions is not the breakage and formation of reaction chemical bands but the binding of the substrate to the enzyme or product release from the enzyme, and hence it cannot be explained by this model (2).

Such consideration has led us to propose a dynamic model of enzyme action (15). My theory is the thermodynamic model for enzyme action. The dynamic model in my theory reveals the logical relation between the internal motion of a protein and enzyme activity. My theory states that many types of protein conformational changes occur along the trajectory of enzymatic reaction, and their properties can be analyzed by protein thermodynamic principles. The transition state of enzymatic reaction is the transition state of enzyme-reactant complex and not the reactant alone. It is the transition state between two different conformational states of enzyme-reactant complex. In contrast, the transition state of a chemical reaction is the transition state of reactants. It clearly shows that behaviors of enzymatic reactions cannot be fully described by traditional transition state theory of chemical reaction. The same enzyme can show different fashions of protein conformational change and transition state when different substrates are utilized. This phenomenon is called substrate-induced switch (32).

We have revised the enzymatic kinetics equation as follows on the basis of the partition function of the protein conformational state (14, 16):

$$K = APe^{-[\Delta E_0 + B_T(T-T_0)]/K_BT} = APDe^{-\Delta E_0/K_BT} \quad (1)$$

where P is the probability of enzyme active conformation, T_0 is the reference temperature, A is the pre-exponential factor at a reference temperature, ΔE_0 is the activation energy of enzymatic reaction at a reference temperature, and D_T is the factor that results from the thermodynamic revision of the transition state of enzymatic reaction.

D_T also represents the dynamic contribution of a protein to enzyme activity because the change in the protein thermodynamic state is induced by a change in protein dynamics.

We have learned from this equation that the activation energy of enzymatic reaction $[\Delta \mathrm{E}_0 + \mathrm{B}_\mathrm{T}(\mathrm{T} - \mathrm{T}_0)]$ is not constant. The change in protein dynamics can alter the activation energy of enzymatic reaction, but not the apparent activation energy $(\Delta \mathrm{E}_0)$ of it.

7. Conclusion

Our analysis indicates that the debate on the dynamic nature of enzyme action has arisen from a misunderstanding of protein thermodynamics and our thinking method. There is a logical relationship between protein thermodynamics and protein function. The protein dynamics takes its role by influencing protein thermodynamic state and thus affect protein function indirectly. Thus, we must seek logic relation between protein function and protein dynamics on the stand of protein thermodynamics (6, 33).

References

1. Frauenfelder, H., Sligar S. G. & Wolynes, P. G. (1991). The energy landscapes and motions of proteins. *Science*. 254, 1598–1603.
2. Zhao, Q. (2013). Nature of Protein Dynamics and Thermodynamics. *Rev. Theor. Sci.* 1, 83–101.
3. Nashine, V. C., Hammes-Schiffer, S. & Benkovic, S. J. (2010). Coupled motions in enzyme catalysis, Current Oponion in Chemical Biology. 14, 644–651.

4. Karplus, M. & McCammon, J. A. (1983). Dynamics of proteins: elements and function. *Ann Rev Biochem.* 53, 263–300.
5. Zhao, Q. (2010). Protein flexibility as biosignal. *Crit. Rev. Eukaryot. Gene. Expr.* 20, 71–84.
6. Zhao, Q. (2013). A Molecular and Biophysical Model of the Biosignal. *Quantum Matter.* 2, 9–16.
7. Cannon, W. R. Singleton, S. F. & Benkovic, S. J. (1996). A perspective on biological catalysis. *Nat Struct Biol.* 3, 821–833.
8. Kohen, A. & Klinman, J. P. (1999). Hydrogen tunneling in biology. *Chem Biol.* 6, 191–198.
9. Daniel, R. M., Dunn, R. V., Finney, J. L. & Smith, J. C. (2003). The role of dynamics in enzyme activity. *Ann Rev Phys Biomol Struc.* 32, 69–92.
10. Kamerlin, S. C. L. & Warshel, A. (2010). At the Dawn of the 21st Century: Is Dynamics the Missing Link for Understanding Enzyme Catalysis? *Proteins.* 78, 1339–1375.
11. Bhabha, G., Lee, J., Ekiert, D. C., Gam, J., Wilson, I. A., Dyson, H. J., Benkovic, S. J. & Wright, P. E. (2011). A dynamic knockout reveals that conformational fluctuations influence the chemical step of enzyme catalysis, *Science.* 332, 234–238.
12. Adamczyk, A. J., Cao, J., Kamerlin, S. C. & Warshel, A. (2011). Catalysis by dihydrofolate reductase and other enzymes arises from electrostatic preorganization, not conformational motions. *Proc Natl Acad Sci USA.* 108, 14115–20.
13. Zhao, Q. (2009). Protein thermodynamic structure. *IUBMB Life.* 61, 600–606.
14. Zhao, Q. (2012). Partition function of protein conformational state. *J. Comput. Theor. Nanosci.* 9, 745–751.
15. Zhao, Q. (2011). Dynamic model of enzymes action. *Protein Pept Lett.* 18, 92–99.
16. Zhao, Q. (2015). A thermodynamic and theoretical view for enzyme regulation, biochemistry (Moscow). 80, 1–7.
17. Henzler-Wildman, K. A., Lei, M., Thai, V., Kerns, S. J., Karplus, M. & Kern, D. (2007). A hierarchy of timescales in protein dynamics is linked to enzyme catalysis. *Nature.* 450, 913–916.
18. Nakagawa, N. & Payrard, M. (2006). The inherent structure landscape of a protein. *Proc. Natl. Acad. Sci. USA.* 103, 5279–5284
19. Pace, C. N. & Tanford, C. (1968). Thermodynamics of the unfolding of β-lactoglobulin A in aqueous urea solutions between 5 and 55, *Biochemistry.* 7, 198–208.
20. Koshland, D. E. (1958). Application of a theory of enzyme specificity to protein synthesis, *Proc. Natl. Acad. Sci. USA.* 17, 1145–1150.

21. Koshland, D. E. (1959). Enzyme flexibility and enzyme action. *J. Cell. Comp. Physiol.* 54, 245–258
22. Teilum, K.,Olsen, J. G. & Kragelund, B. B. (2009). Functional aspects of protein flexibility. *Cell Mol Life Sci.* 66, 2231–47.
23. Kokkinidis, M., Glykos, N. M. & Fadouloglou, V. E. (2012). Protein flexibility and enzymatic catalysis. *Adv Protein Chem Struct Biol.* 87, 181–218.
24. Zavodszky, P., Kardos, J., Svingor, A. & Petsko, G. A. (1998). Adjustment of conformational flexibility is a key event in the thermal adaptation of proteins, *Proc. Natl Acad. Sci. USA*. 95, 7406–7411.
25. Austin, R. H., Beeson, K. W., Eisenstein, L., Frauenfelder, H. & Gunsalus, I. C. (1979). Dynamics of ligand binding to myoglobin. *Biochemistry.* 14, 5355–5373.
26. Tsou, C. L. (1993). Conformational Flexibility of Enzyme Active Sites, *Science.* 262, 380–381.
27. Lonhienne, T., Gerday, C. & Feller, G. (2000). Psychrophilic enzymes: revisiting the thermodynamic parameters of activation may explain local flexibility, *Biochim. Biophys. Acta.* 1543, 1–10.
28. Klinman, J. P. (2010). Control of Active-Site Compression. *Nat Chem (News & Views).* 2, 907–909.
29. Klinman, J. P. (2009). An integrated model for enzyme catalysis emerges from studies of hydrogen tunneling. *Chem Phys Lett.* 471, 179–193.
30. Michaelis, L. & Menten, M. L. (1913). "Die Kinetik der Invertinwirkung". *Biochem Z* 49, 333–369.
31. Pauling, L. (1946). Molecular Architecture and Biological Reactions. *Chem. Eng. News.* 24, 1375–1377.
32. Simone, D. *et al.* (2004). A Substrate-induced Switch in the Reaction Mechanism of a Thermophilic Esterase. *J. Biol. Chem.* 279, 6815–6823.
33. Zhao, Q. (2013). Allodynamic Regulation of Protein Activity. *Quantum Matter.* 2, 144–152.

Chapter 16

Protein Ligand Interaction

Summary: The protein ligand interaction is an important process in biology and is connected to the biological regulation of protein and enzyme. The binding between enzyme and substrate, protein-protein interaction, ligand-receptor are examples of this process. Here we analyze the basic mechanism for it on the basis of the protein conformation concept.

Protein-ligand interaction, including substrate-enzyme binding, agonist-receptor binding, regulator-receptor binding, is an important process in biochemistry. According to the experimental method and working mechanism, it can be classified as ligand migration and ligand protein interaction. The ligand migration occurs when a small ligand, such as oxygen, CO, He, which can bind to many sites of a protein, binds to a protein (1). Ligand migration mechanism is normally studied by biophysicists. The ligand protein interaction occurs when there is specific interaction between protein and ligand, normally, involving a protein that has one or more sites which can bind to a ligand specifically. Much knowledge of biochemistry has come from the studying of protein ligand interaction.

1. Ligand Migration Within a Protein

A protein has many internal cavities, several hydrophobic surfaces. Small ligand, such as Xe and CO, can occupy such sites and bind to a protein weakly. A molecule can occupy any of these sites and may migrate from one site to another. The process is called ligand migration. It was first observed in the binding between hemoglobin and CO. Several important conclusions have been obtained from this approach and we summarize it here (1–8).

Firstly, ligand migration is not the simple thermal process of diffusion of a ligand, for it has relation to protein dynamics. In other words, ligand migration is accompanied by the protein conformational transition.

Secondly, a protein has multiple conformational states at a similar energetic level.

Thirdly, the protein conformational state is changeable under different conditions.

Fourthly, the ligand migration is influenced by other molecules, for other ligands can competitively bind to a same site.

The process of ligand migration could be expressed in Figure 1.

For the migration of larger molecules, such as NAD, global conformation of a protein is needed and it often includes domain motions or thermal reorganization of subunits of a protein complex. The protein ligand interaction could induce water structure reorganization, where by there is a binding of protein and water, and then the protein

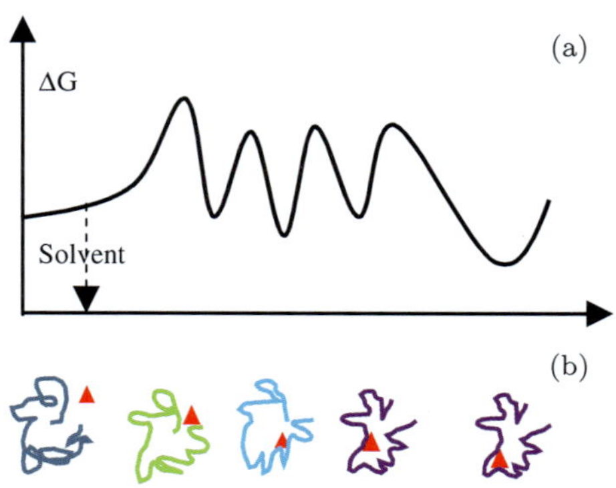

Figure 1. Diagram for ligand migration within a protein.

Figure 1a shows the energetic feature of ligand migration along a migration pathway. Figure 1b shows an illustration of protein ligand migration. The red triangle represents ligand. For the migration of CO or other small molecules, the local protein conformational change is needed.

conformation reorganization includes water structure reorganization at the binding site of protein ligand interaction (9).

2. The Nature of Protein Ligand Interaction

In ligand migration mechanism, the ligand can bind to many sites of a protein non-specifically. Normally, the term protein ligand interaction refers to specific interactions between a ligand and a protein, such as binding between enzyme-substrate. This type of protein ligand interaction can be described by the laws of common chemical reaction.

Thermodynamics of protein ligand interaction: normally, the protein ligand interaction can be described by the law of equilibrium thermodynamics as follow:

$$P + L = PL$$

$$K_a = \frac{C_{PL}}{C_P C_L}$$

$$\Delta G^\circ = -RT\ln K_a$$

$$\Delta G^\circ = \Delta H^\circ - T\Delta S^\circ$$

Where Ka is association constant, C_{PL} is the concentration of protein-ligand complex, C_P is the protein concentration, and C_L is the concentration of ligand. ΔG° is the difference of Gibbs free energy for a given reaction at reference condition, ΔH° is enthalpy change at reference condition, ΔS° is the entropy change of the reaction at reference condition.

This expression can give a satisfactory explanation for the protein-ligand interaction on the basis of equilibrium thermodynamics. However, it is unable to deal with the dynamic events of protein ligand interaction for they are laws of equilibrium thermodynamics.

Dynamic events of protein ligand interaction: theoretically speaking, the conclusions obtained in ligand migration are applicable to the cases of protein ligand binding, but revisions should be made. The main difference is that the migration of larger molecules, such as the substrate of enzymes, is completely controlled by protein conformational change and effect of diffusion can be ignored in substrate enzyme binding. In fact, the flexible protein recognition model (FPRM) is the revision of ligand migration mechanism (10). According to the protein thermodynamic structure theory, we revise FPRM model as follows:

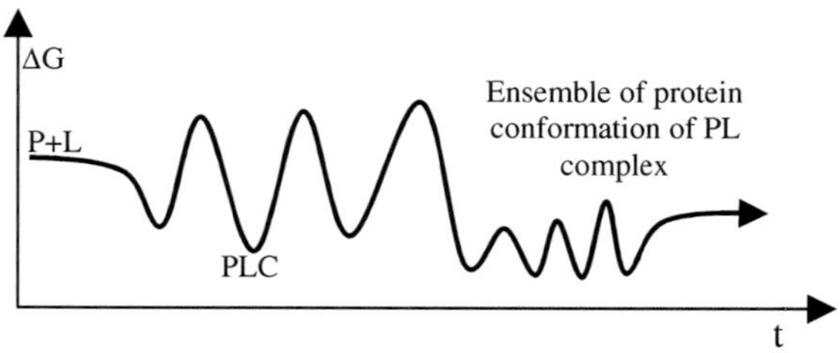

Figure 2. Diagram for protein ligand interaction.

P represents protein, L represent ligand, PLC represent protein ligand complex, and t is time.

Briefly speaking, the protein ligand binding is initiated and driven by a long range force between protein and ligand. After protein and ligand have encountered each other, a specific interaction (Fig. 2) between them occurs and a protein-ligand complex is formed. The complex of protein ligand then undergoes several steps of protein conformational change at global level, ending with several equilibrium states of protein-ligand complex or ensembles of protein conformations of protein-ligand complexes. Several properties of a protein ligand complex should be addressed. 1) In view of biochemistry, a conformation at global level, such as active and inactive of enzyme or receptor, may predominate under a condition; 2) the predominating conformation of a protein ligand complex is changeable under different conditions; 3) any conformations at global level comprised of many conformers at atom-level.

Protein ligand interaction changes the flexibility of a protein at some sites. The protein ligand interaction alters the protein dynamics of some residues at different sites of a protein, but not all of the residues of a protein. This property can be studied by several techniques, such as NMR, or the hydrogen exchange technique.

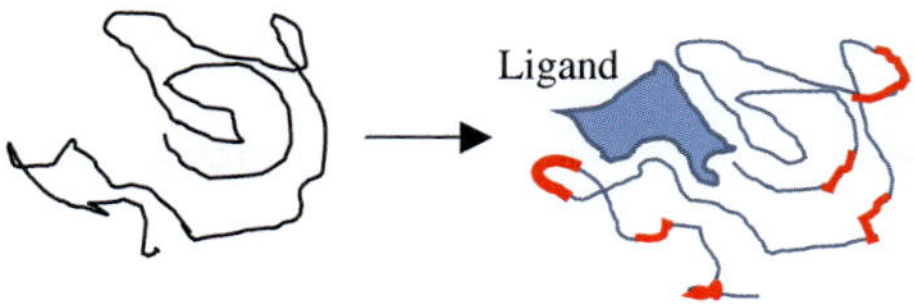

Figure 3. Diagram of protein flexibility change in protein ligand interaction. When a protein binds to a ligand, it can induce a change in protein flexibility at many areas of the protein. Red areas denote such sites or area in which a major change of protein flexibility is produced.

The interaction between the protein and ligand can induce a change in protein thermodynamics. But the change of protein flexibility (Fig. 3) varies from site to site. The change in protein flexibility at one site may be larger and it may be smaller at the other site (11). Normally, the protein ligand interaction can only induce a remarkable change in protein flexibility at limited sites.

3. The Entropy–Entropy Compensation and the Timescale of Protein Ligand Scale

As we have discussed, the protein ligand interaction was initiated by the chemical affinity between protein and ligand, and is processed or driven further by protein thermodynamic reorganization or protein conformational change. Unlike typical and simple chemical reactions which are driven by chemical potential, the protein ligand interaction shows differing features. A lot of studies on thermodynamic interactions between protein and ligand have been processed, much data has been obtained and concept of entropy–enthalpy compensation in protein ligand interaction has been well-established. When larger amount data were pointed in the coordinate system of Figure 4, they were all located in dark area in Figure 4 (12, 13). All points in the line of Figure 4 has the same Gibbs free energy of protein ligand binding.

From this figure, we can learn that:

$$\Delta H \approx T\Delta S$$

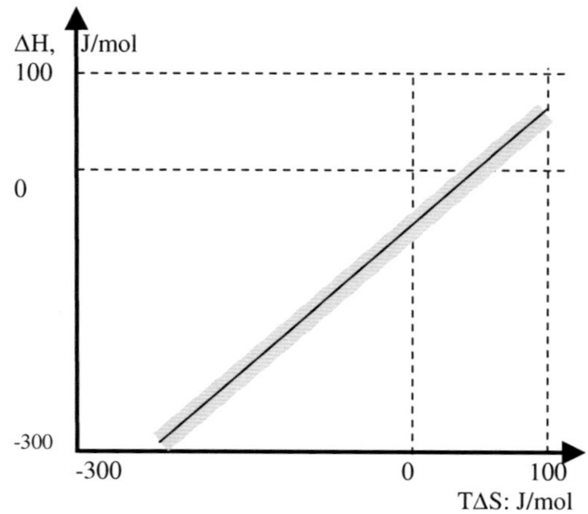

Figure 4. Diagram of Entropy–enthalpy compensation. The data can be seen in many papers. This is a summation.

Or

$$\frac{d(\Delta H)}{d(T\Delta S)} \approx \text{constant}$$

This equation is called entropy–entropy compensation. It states that when a condition is altered, a change in the enthalpy of protein ligand interaction can be compensated in some degree by an opposite change of entropy, where the binding affinity of protein ligand remains a constant.

Several important conclusions could be deduced from the entropy–entropy compensation phenomenon:

First, the thermodynamic nature of protein ligand interaction is changeable under different conditions. It is also a powerful evidence for protein flexibility.

Second, there are many protein thermodynamic states within a protein which are connected to protein ligand interaction; or in other words, a protein ligand complex shows many different conformations under different conditions and energy of these conformers is at a similar level. This agrees with the results obtained in ligand migration tests.

According to irreversible thermodynamics and protein thermodynamic structure theory (14, 15), we can obtain:

$$T = S_E/v$$

$$S_E = W$$

$$S = \text{klin}W$$

Where S_E is the protein conformation ensemble or sample space, v is the transition rate. W is the sample space of thermodynamic state. K is Boltzmann constant.

For any interaction of protein and ligand, one point can be found and there is a following equation

$$S \propto -\Delta G$$

Then we can obtain

$$T \propto e^{a(-\Delta G)} \quad (1)$$

Where a is constant.

This equation indicates that the time scale of protein ligand interaction is connected to its binding affinity. The closer the affinity between protein and ligand, the longer the time needed for its interaction. This nature differs from that of simple chemical reactions in principle greatly. For simple chemical reactions, the chemical rate is determined by chemical potential and the chemical reaction is processed rapidly for high chemical potential, and vice versa.

The timescale of protein ligand interaction has been well summarized and it can be seen in Figure 6 of Chapter 14. We can see that our conclusion agrees well with the experimental data.

4. The Ligand Receptor Binding Changes the Distribution Curve of the Active Conformation of a Receptor or Enzyme

The ligand-protein complex represents a new species of proteins, thus the distribution curve of the active conformation of substrate-enzyme complex can be measured (16, 17). In solution, the active conformation curve of a native enzyme cannot be measured for it is formed in the process of substrate and enzyme binding.

By comparing distribution curves of different substrate-enzyme complexes, the shift of protein active conformation curves can be observed. The conformational shift was also observable when comparing active conformation curves of substrate enzyme complexes when the enzyme works in different conditions, such as in the presence of urea or other protein denaturants.

According to the partition function of the protein conformational state, we have drawn the conformational shift in Figure 5.

The protein conformation isolated by biochemical method or enzymatic kinetics is at a global level and in fact represents an ensemble of protein conformations at atom level. This is the thermodynamic mechanism for the shift of population of protein conformation.

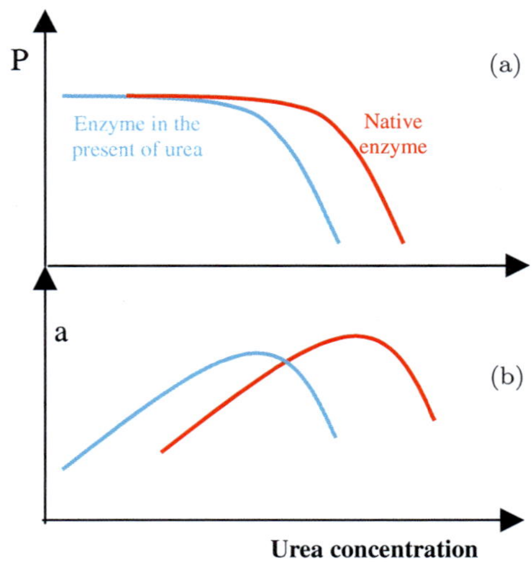

Figure 5. Diagram showing the shift of the distribution curve of active conformations of an enzyme substrate complex.

P is the probability of active conformation of enzyme substrate complex; a is the enzyme activity. Figure 5a is the distribution curve of active conformation of enzyme. Figure 5b is the profile or activity curve of enzyme.

5. Thermodynamic Reorganization Model for Protein Ligand Interaction

In retrospect: Two different models, namely the induced-fit model or conformational selection model, are dominating our understanding of protein ligand interaction (18–19).

The induced-fit model, proposed by DE Koshland, is the revision of the key-lock model. It states that the shape or conformation of a protein can undergo reversible change in some degree and that the binding to a ligand can induce a change of conformation or shape of a protein. The most favorable conformation of a protein, which binds to the ligand tightly, is the product of protein ligand interaction. There is a sequential change in protein conformation when protein ligand interaction occurs. The rightness of the induced–fit theory has been confirmed by numerous experimental tests and it has dominated our understanding about protein ligand interaction for nearly 50 years.

Early in 1964, a Russian scientist, Straub, had proposed a conformation selection hypothesis for enzyme substrate binding (20). It states that in solution, a protein shows many different conformations and that the ligand can selectively bind to one conformation of the protein. But at times, this hypothesis did not attract any attention from the scientific community. In the 1990's, this theory was re-mentioned by different groups of USA scientists who were studying protein dynamics, and thus it was gradually accepted by some scientists over the world.

The major difference between the two models is that within the conformational selection model, protein conformation, which binds to a ligand exists spontaneously in the absence of ligands, and within induced fit model, the protein conformation which binds to a substrate was generated in the interacting process between substrate and enzymes. Another difference is that the protein conformational change is an intrinsic property of a protein within the selection model, and the protein conformational change is induced by the interaction between the enzyme and substrate within the induced-fit model. Although the protein conformational selection model has prevailed over the world, it is not a mature theory.

The studying of protein thermodynamic structure theory has resulted in a new theory for protein ligand interaction, namely thermodynamic reorganization. We will discuss the advantages and theoretical flaws of the two models by applying the principles of our theory.

Firstly, according to the partition function of protein conformational state, some types of protein conformations, such as active conformation, does not exist in solution when the ligand is absent in protein solution. In fact, by reading the researches in favor of the population hypothesis, we can find that many study the protein conformation at the atom or local level. The difference between the global and local conformation has been ignored, however.

Secondly, although the protein ligand interaction can change the flexibility of the protein at some sites, the overall profile of protein flexibility or dynamics usually does not change. The rate of protein conformational change or the dynamic nature of protein has not taken into account in the theory of induced-fit, while the conformation selection model and our model can however give a satisfactory explanation to this feature.

References

1. Ostermann, A., Waschipky, R., Parak, F. G., & Nienhaus, G. U. (2000). Ligand binding and conformational motions in myoglobin. *Nature*, 404(6774), 205–208.
2. Tian, W. D., Sage, J. T., & Champion, P. M. (1993). Investigations of ligand association and dissociation rates in the "open" and "closed" states of myoglobin. *Journal of molecular biology*, 233(1), 155–166.
3. Šrajer, V., Ren, Z., Teng, T. Y., Schmidt, M., *et al.* (2001). Protein conformational relaxation and ligand migration in myoglobin: a nanosecond to millisecond molecular movie from time-resolved Laue X-ray diffraction. *Biochemistry*, 40(46), 13802–13815.
4. Young, R. D., & Bowne, S. F. (1984). Conformational substates and barrier height distributions in ligand binding to heme proteins. *The Journal of chemical physics*, 81(8), 3730–3737.
5. Frauenfelder, H., Sligar, S. G., & Wolynes, P. G. (1991). The energy landscapes and motions of proteins. *Science*, 254(5038), 1598–1603.

6. Frauenfelder, H., Parak, F., & Young, R. D. (1988). Conformational substates in proteins. *Annual review of biophysics and biophysical chemistry*, 17(1), 451–479.
7. Tetreau, C., Blouquit, Y., Novikov, E., Quiniou, E., & Lavalette, D. (2004). Competition with xenon elicits ligand migration and escape pathways in myoglobin. *Biophysical journal*, 86(1), 435–447.
8. Young, R. D. & Bowne, S. F. (1984). Conformational substates and barrier height distributions in ligand binding to heme proteins. *The Journal of chemical physics*, 81(8), 3730–3737.
9. Levy, Y. & Onuchic, J. N. (2006). Water mediation in protein folding and molecular recognition. *Annu. Rev. Biophys. Biomol. Struct.*, 35, 389–415.
10. Grünberg, R., Leckner, J., & Nilges, M. (2004). Complementarity of structure ensembles in protein-protein binding. *Structure*, 12(12), 2125–2136.
11. Bonvin, A. M. (2006). Flexible protein–protein docking. *Current opinion in structural biology*, 16(2), 194–200.
12. Chodera, J. D. & Mobley, D. L. (2013). Entropy-enthalpy compensation: role and ramifications in biomolecular ligand recognition and design. *Annual review of biophysics*, 42, 121.
13. Teilum, K., Olsen, J. G., & Kragelund, B. B. (2009). Functional aspects of protein flexibility. *Cellular and Molecular Life Sciences*, 66(14), 2231–2247.
14. Zhao, Q. (2013). Nature of protein dynamics and thermodynamics. *Reviews in Theoretical Science*, 1(1), 83–101.
15. Zhao, Q. (2009). Protein thermodynamic structure. *IUBMB life*, 61(6), 600–606.
16. Zhao, Q. (2012). Partition Function of the Protein Conformational State. *Journal of Computational and Theoretical Nanoscience*, 9(5), 745–751.
17. Zhao, Q. (2015). A thermodynamic and theoretical view for enzyme regulation. *Biochemistry (Moscow)*, 80(1), 1–7.
18. Koshland, D. E. (1995). The key-lock theory and the induced fit theory. *Angewandte Chemie International Edition in English*, 33(23–24), 2375–2378.
19. Csermely, P., Palotai, R., & Nussinov, R. (2010). Induced fit, conformational selection and independent dynamic segments: an extended view of binding events. *Trends in biochemical sciences*, 35(10), 539–546.
20. Straub, F. B. & Szabolcsi G. O dinamicseszkij aszpektah sztukturu ··· fermentov. (On the dynamic aspects of protein structure) In: *Molecular Biology*, Problems and Perspectives (Braunstein, A. E., ed.), 1964 pp. 182–187 (in Russian), Izdat. Nauka.

Chapter 17

The Receptor Activation and Fashions of Protein Regulation

Summary: The receptor activation is an important process in biology. The signaling activity of a receptor is initiated at this step. Many types of biological information or signals are integrated in this process, so it has attracted much attention from scientific community in the past. Many valuable opinions about it have been proposed as well. The receptor activation is a good example of protein thermodynamic reorganization and protein regulation. In this chapter, we will show that the protein thermodynamic structure theory can give satisfactory explanations for much of the knowledge we have obtained in this field. In principle, enzyme regulation is same as the receptor activation and we will discuss some general trend of the receptor and enzyme activation.

Receptors are signaling proteins which can respond to a specific agonist or stimulus and transform many types of signals into protein conformational signal. Normally, receptors are on the surface of a cell and integrate diverse types of information of environmental factors which finally initiates signaling activities of a cell (1).

Theoretically, the mechanism of receptor activation and receptor regulation is identical to that of enzyme regulation, but the receptor activation shows unique properties and was studied separately in the past.

The signaling proteins, including receptors, have active conformations with low stability. This endows its high sensibility to a change in the environmental factor, hence the distribution curve of active conformation of receptors is bell-shaped (2–4). For the channels and the receptors, the profiles of their activity curve are similar.

According to the dynamic model for allodynamic regulation of protein activity, all factors, including agonist, antagonist, activator and environmental factors, have an influence on the probability of active conformation of the receptor.

In this chapter, we will discuss the basic principles of receptor activation, the integration mechanism of diversified factors or information in a receptor, and some trends of protein regulation from the perspective of molecular biology.

1. The Two-States Hypothesis for Receptor Activation

The simple model of receptor activation is the two-states model (5). It assumes that a receptor can show two different conformations, specifically the active and inactive conformation, under different conditions. Receptor activation is achieved by the transformation from inactive conformation to active conformation of a receptor. The protein conformational change needed for receptor activation occurs when it binds to its agonist or is provoked under stimulus. This principle is shown in Figure 1.

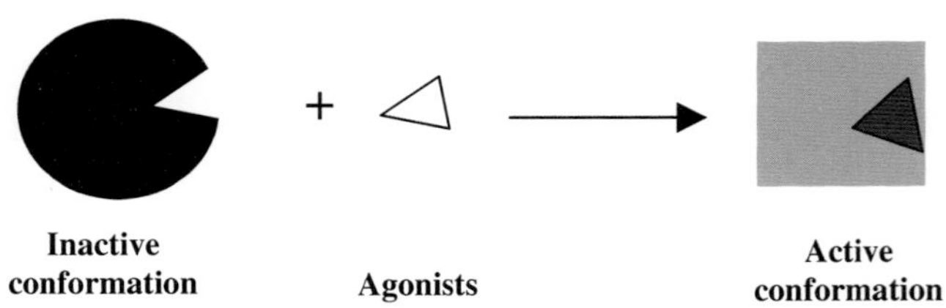

Figure 1. Diagram for receptor activation.

When inactive conformation of a receptor encountered a ligand, it can change its shape and the active conformation of the receptor is produced.

The binding between the agonist and the receptor can be described by the laws of thermodynamics and has been discussed in the chapter of protein ligand interaction.

Although this hypothesis is simple, it is the theoretical foundation for receptor activation and other approaches that are developed based on the concepts proposed within this model.

The theoretical flaw of the two states hypothesis for the receptor activation or the protein regulation is obvious and several aspects of receptor activation have not been addressed within this model. We will discuss these problems as follows.

2. The Conformations of Protein-Agonists Interaction

A protein, or any part of it, shows numerous types of conformations. The receptor agonist complex can be considered a new species of a protein. The conformation of protein-agonist complex is changeable, which is related to the receptor activation and the receptor activity.

The deviation of the protein conformation of a receptor-agonist complex from standard or diversified conformations of receptor–agonists can be clearly seen in the following: 1) the competitive antagonist and agonist bind to the same site of a receptor, but the competitive antagonist shows no potency to trigger receptor activity; 2) different agonists bind to a receptor with different affinities and show different potencies; 3) the potency of an agonist is changeable under different conditions.

The binding of a receptor to agonist (or antagonist) is shown in Figure 2.

Thus, one conclusion can be obtained, which is that the receptor and agonist can bind via different conformations of a receptor.

The distribution curve of the active state (conformation) of a GABA receptor against urea concentration is shown in Figure 3.

In Figure 3, we can see that the energetic level of active conformation of a receptor is changeable, thus the conformation of protein-agonist binding is also changeable. A substance can act as a fully agonist in one condition, a partial agonist in another condition, and

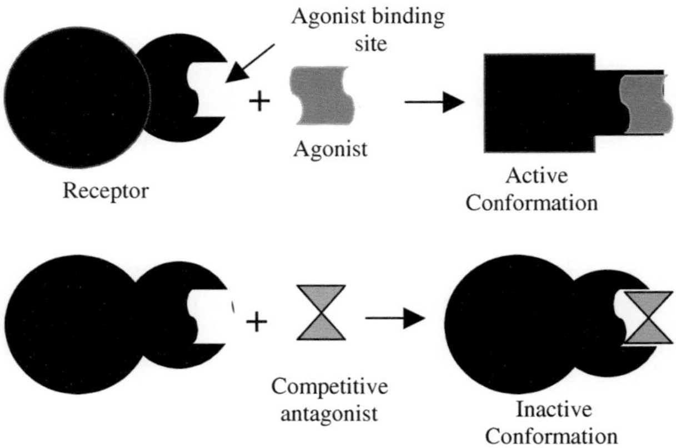

Figure 2. The diagram of binding between agonist (competitive antagonist) and the receptor.

The binding of the competitive agonist does not result in active conformation of the receptor. The binding of an agonist to a receptor induces the generation of active conformation. It shows that the agonist binding site of the receptor can show different conformations.

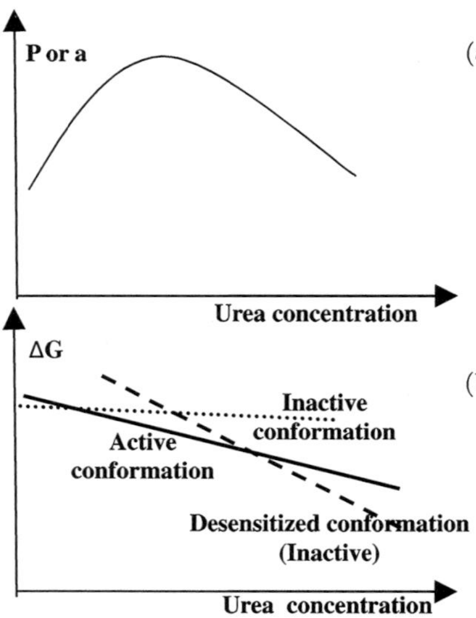

Figure 3. The distribution curve of active conformation of the GABA receptor and the stability curve of conformations.

Figure 3a is drawn up according to the data of Ref (6). The stability curves (Figure 3b) are drawn according to the partition function of protein conformational state (4) (it is non-empirical data).

an antagonist under conditions far away of the optimum condition for receptor activity.

3. Diversified Conformations and the Thermodynamic Structure of a Receptor

By applying diverse methods, many types of protein conformations of a receptor were tested experimentally (7, 8). In order to give a satisfactory explanation of the experimental results, it is necessary to define the standard (or ideal) protein thermodynamic structure of a receptor which is shown in Figure 4.

According to principles of protein thermodynamic structure, we have formulated the standard fashion of protein thermodynamic structure of a receptor as follows.

1. Diversified conformations of a receptor show a hierarchical structure and the active conformation of a receptor is at upper level while other types of conformations are at fundamental level.
2. There is no interaction of conformations at fundamental levels. In other words, the interaction between the agonist (or regulating

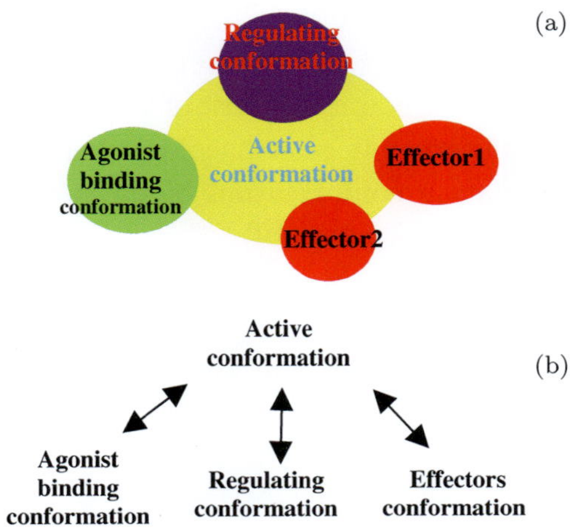

Figure 4. The standard fashion of Protein thermodynamic structure of a receptor.

Figure 4a is an illustration of the protein thermodynamic structure of a receptor. Figure 4b shows the hierarchical structure of it.

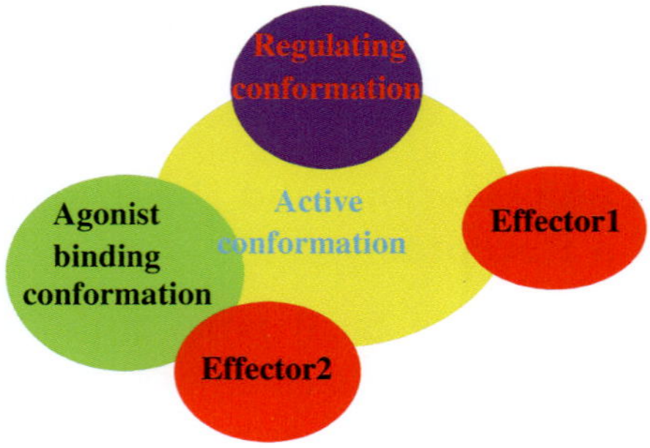

Figure 5. The Biased agonist.

In this fashion of receptor activation, the agonist binding conformation can influence the effector 2 conformation in two different ways: the first way is via active conformation of the receptor, the second is through direct interaction.

conformation) and the active conformation of a receptor is permitted, but the direct interaction between the agonist binding conformation and the effector conformation is not permitted.

3. The agonist, a noncompetitive regulator has an influence on the probability of active conformation of a receptor in solution.
4. Different agonists can stimulate the activity of the effector 1 and the effector 2 with no selection.

These principles can account for most of the results of the receptor activation we have observed in pharmacology. For G protein coupled receptor, the dominating conformation of the active state is seven transmembrane (7–TM) α-helices (9).

In fact, the trend of protein conformation of a receptor is more complex. If there were direct interactions between agonist binding conformations, the biased agonist would be observed (10).

4. The Allosteric Regulator

Many drugs act as receptor regulators and bind to the receptor on the opposite side of the agonist. For the most of them, they are allosteric regulators and bindings between them and the receptor are independent of the receptor-agonist binding. The antagonist binding potherse (or conformation) is independent of that of the agonist-receptor conformation.

According to the thermodynamic model for protein regulation, the regulator can alter equilibrium between the inactive and the active state of the receptor. It is therefore unnecessary to interfere the binding ability between the agonist and the receptor.

Theoretically, gene mutation and protein modification (phosphorylation) are special allosteric regulators of a receptor.

5. Environmental Factors as a Regulator of a Receptor

Receptor activity is influenced by many environmental factors. Unlike the agonist of a receptor which targets one or a few specific receptors, environmental factors, which include mechanical forces, pressure, protein denaturant, temperature, pH, salt concentration and species of the working buffer etc. have great impact on the activities of all receptors, or proteins. Their effects on receptor activity can be seen in many specific ways.

Firstly, some receptors are altered in biological evolution, which make them sensitive to one environmental factor, as observed in the case of the voltage-sensitive channel. In this case, the environmental factor is specific and this response is functional in physiology.

Secondly, the environmental factors can regulate protein conformation or the conformational change of a receptor and influence the receptor activities indirectly. Although some types of environmental factors, such as the salt of buffer, may not trigger the signaling activity of a receptor, they can regulate the receptor activity by influencing the probability of receptor active conformation. It is well-known that a change in temperature can influence the activities of all receptors.

The effect of environmental factors on receptor activities can be analyzed via the distribution curve of active conformation of the receptor in the presence of agonist (or protein-agonist complex). As we can see in Figure–3B, urea impacts receptor activity at 0–2 Mol. In this case, urea acts as an environmental factor.

As little information on the distribution curve of an active conformation of receptors exists currently, it should be carefully studied in future.

6. An Integration of Diversified Types of Information Within a Receptor

All types of information are integrated within a thermodynamic receptor and the final output or the activity of a receptor results from this type of integration. From the perspective of thermodynamics, all the factors have a joint impact on receptor activity via the same mechanism: influencing the probability of an active conformation. But from the view of experimental science or pharmacology, the integration of diversified types of information or the signal of a receptor can be classified as follows.

The Agonist-antagonist-receptor interaction: This type of integration of information has been well studied previously in field of pharmacology. The agonist and antagonist can both influence the probability of active conformation of the receptor. An agonist can act as an antagonist under different conditions, and vice versa (11).

The Agonist-environmental factors interaction: the receptor conformation can be influenced by environmental factors. Their effects can be judged osing the distribution curve of the active conformation of the receptor (2–4).

The Gene mutation of a receptor and the interaction among them. The expression of the active conformation of a receptor is reliant on the intrinsic nature of a receptor, while gene mutation has a great impact on all properties of a protein and receptor. The constitutive mutants of a receptor, which show the activities under the condition that the wild receptors show no activity, have been discovered, but the studies on this topic are just beginning and we need more data to make further progress. The cooperation and neutralization among effects of gene mutation should be observed according to the protein thermodynamic structure theory. It has been shown that the mutations of the receptor with constitutive activities are located in many areas of it (12).

The Protein modification and its effects. The protein modification, particular to phosphorylation, is a common mechanism for receptor regulation (13). Most phosphorylation sites of proteins are in flexible areas of receptors. Like gene mutations, the protein modification can alter many properties of a receptor. It can activate or desensitize a receptor under different conditions.

7. The Distribution Curve of an Active Conformation of a Receptor

The fashion of bio-information integration within a receptor can be analyzed using the distribution curve of an active conformation of a receptor (2–4). Here we will discuss the working mechanism of an inverse agonist. In order to analyze its working mechanism, we need to examine the distribution curves of active conformations of wild receptors, the protein of complex of agonist-wild-receptor, the mutant receptor with constitutive activities, and the complex of agonist-mutant-receptor. In Figure 6, we have drawn the basic profile of the inverse agonist and its receptor (it is a theoretical explanation or prediction for there is not enough data for it).

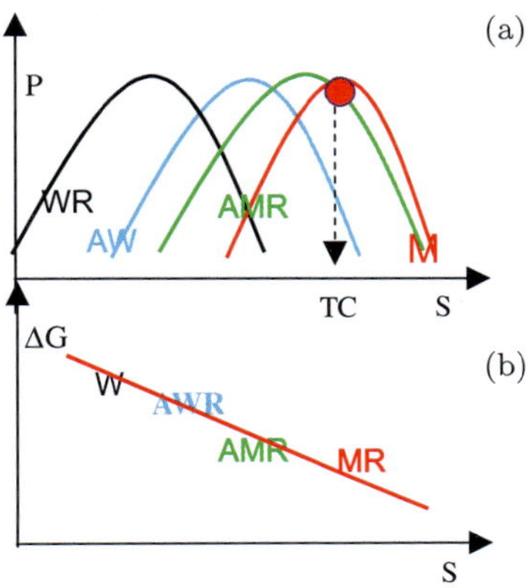

Figure 6. The distribution curves of the active conformation of a receptor.

Where WR is the wild receptor, AWR is the complex of agonist-wild-receptor, MR is the mutant receptor with constitutive activities, and AMR is the complex of the agonist-mutant receptor. TC represents the physiological condition or the test condition for a receptor. S is the variable of the environment which can act as a receptor, such as the temperature, voltage and concentration of urea. Figure 6a shows the distribution curve of the active conformation of a receptor complex. Figure 6b represents the stability of the active conformation of the diversified complex of a receptor (a more precise analysis will be established on the stability curves of the active conformation of receptors, but it will not be discussed in detail here).

In this Figure, we can see that the optimum conditions of the diversified receptors are different from each other. The mutant receptor with constitutive activities has the lowest stability. Both the agonist and the gene mutation of a receptor are the factors which influence the stability of the active conformation of a receptor, and their effects can either be neutralized or cooperate under some conditions. It has been reported that the protein flexibility of a receptor is related to the working mechanism of inverse agonists. This discovery is in accordance with our theory (14, 15).

8. Some Types of Receptor Activation

According to the model proposed in this chapter, we have predicted several types of receptor activations that have not been validated.

The decoupling mechanisms for receptor inactivation: if the coupling between the agonist binding conformation and active conformation of a receptor is lost, the receptor will not be activated by the agonist. The principle is shown in Figure 7.

Two scientific criteria can be used to identify it: 1) when a receptor bind to its agonist, it shows no activity; 2) the binding of agonist to the receptor cannot induce the conformational change at the other area, but binding between the regulators and the receptor can induce a conformational change of the receptor. It may appear under the condition that a protein has a low stability, such as at the near denaturation state, or when the mutants are present.

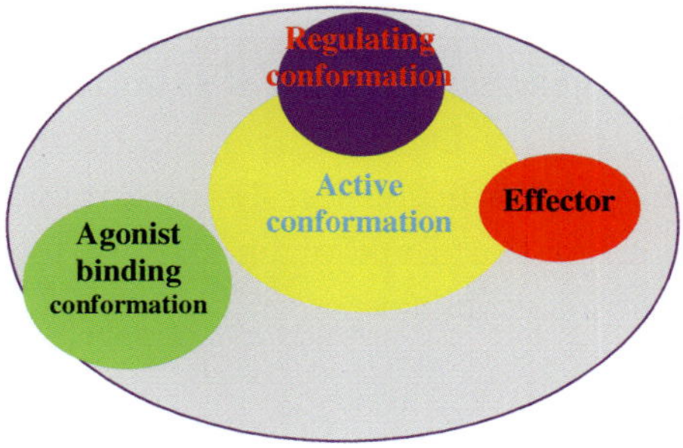

Figure 7. The decoupling between the agonist binding conformation and the active conformation of a receptor.

In this case, the agonist binding conformation and the active conformation of a receptor become independent completely. The change of the agonist binding conformation thus cannot induce a change at the active conformation of the receptor.

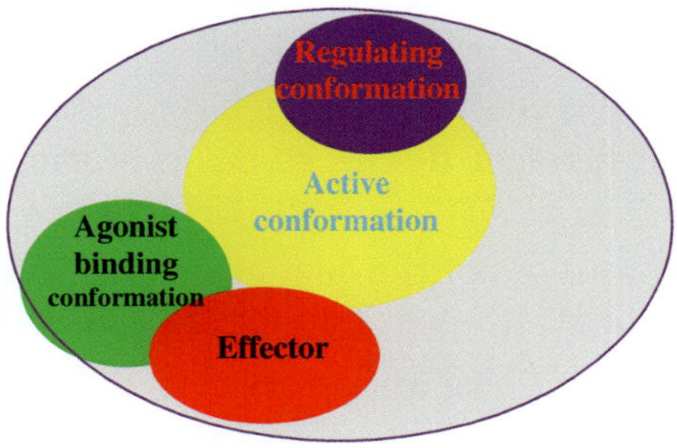

Figure 8. The de-regulating receptor.

In summary, this chapter has provided a new a way to understand receptor activation thermodynamically. More data and research is needed in the future for further progress in this area.

The de-regulating receptor: if the direct coupling between the agonist binding conformation and the effector conformation become very strong, while coupling between the active conformation and the effector become weaker, the receptor will deregulate, and become insensitive to the regulator (or the noncompetitive antagonist). The fashion of its protein thermodynamic structure is shown in Figure 8. It must be pointed out that this type of receptor should be eliminated in biological evolution. The criteria for it is as follows: 1) the receptor activation is sensitive to agonists, but not noncompetitive antagonists; 2) The regulator or noncompetitive agonist still has an ability to bind to a receptor, however without any impact on the receptor activities; 3) the dominating structure of active conformation, or the dominating structure of it, has nothing to do with the receptor activities.

9. The Regulation Fashions of the Ribonucleotide Reductase

The Ribonucleotide reductase (RNR) catalyzes the formation of deoxyribonucleotides from ribonucleotides and is essential for the maintenance of cellular physiology (16, 17). The activity regulation of this enzyme contributes much to the viability of bacteria. The regulation trend of Ribonucleotide reductase has been well studied and it provides an excellent example for protein regulation. It has been observed that two different mechanisms have been utilized for enzyme regulation. The first refers to the global conformational change of Ribonucleotide reductase. When ATP is bound to the enzyme, the active conformation ($\alpha_2\beta_2$)of the enzyme is predominant in solution; when dATP is bound to the enzyme, the inactive conformation ($\alpha_4\beta_4$) is predominant. Thus, we can see that the ratio of ATP/dATP determines the abundance of active conformation of enzyme and thus controls enzyme activity indirectly. The different conformations in living condition arising from binding of Ribonucleotide reductase to ATP/dATP can serve as a powerful evidence for our theory.

The second mechanism refers to the local conformations of a regulator binding site of Ribonucleotide reductase. It has shown that dTTP, dATP, dGTP, dCTP can bind to a site of Ribonucleotide reductase and this regulates the substrate specificity of the enzyme. In other words, when the enzyme is bound to the regulator dTTP, the enzyme is prone to utilizing the substrate GDP; the binding of dATP makes enzyme use substrates of CDP and UDP; the binding of dGTP make enzyme use substrate ADP. The underlying reason is that the conformations of substrate and regulator are coupling with each other.

Many types of protein conformation have involved in the enzyme regulation of Ribonucleotide reductase and they are depicted in Figure 9.

In Figure 9, we see that regulator (or specificity effector) binding conformation of the enzyme is independent to global conformation (or active conformation) of the enzyme and it can structure itself to some degree; and thence many different conformations can be formed at this site (Figure 9c). The conformational coupling between the

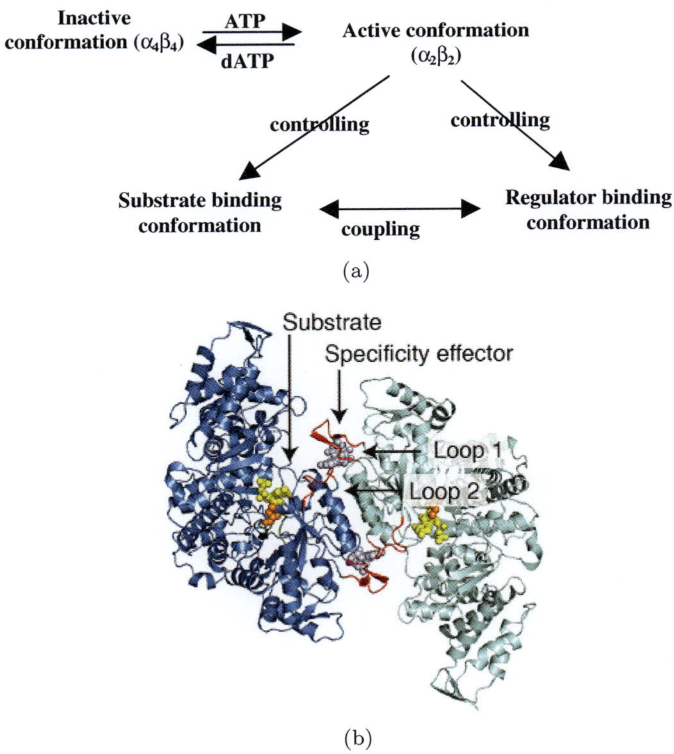

Figure 9. The regulation fashions of Ribonucleotide reductase.

Figure 9a: The protein thermodynamic structure of Ribonucleotide reductase. Figure 9b: The active conformation of Ribonucleotide reductase. Figure 9c: The local conformation at substrate and the regulator binding site of Ribonucleotide reductase. (Figure 9b and 9c was a reproduction from Ref. 17, with permission).

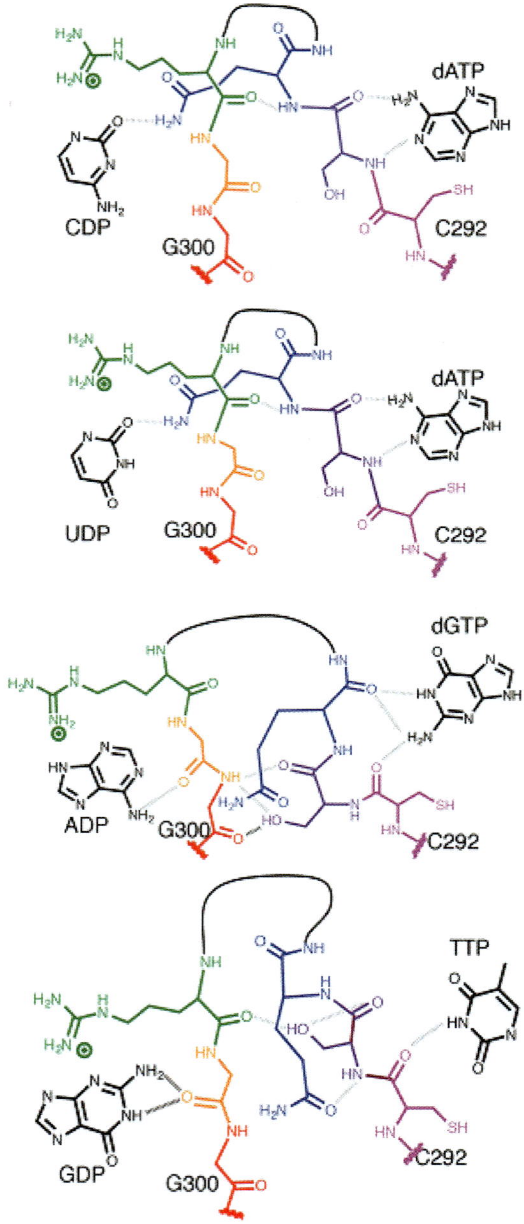

(c)

Figure 9. (*Continued*)

regulator binding conformation and the substrate binding conformation occurs and it regulates the specificity of enzyme substrates. The enzyme activity regulation is mainly regulated by global conformational change and is finely regulated by local conformational change. Different protein conformations of the enzyme are well structured and thus enzyme activity can be properly processed according to the physiological requirement of the cell.

References

1. Wrana, J. L., Attisano, L., Wieser, R., Ventura, F., & Massague, J. (1994). Mechanism of activation of the TGF-b receptor. *Nature*, 370(6488), 341–346.
2. Zhao, Q. (2015) A thermodynamic and theoretical view for enzyme regulation, *biochemistry (Moscow)*, 80, 1–7.
3. Zhao, Q. (2013) Allodynamic Regulation of Protein Activity. *Quantum Matter*, 2, 144–152
4. Zhao, Q. (2012) Partition function of protein conformational state. *J. Comput. Theor. Nanosci.* 9, 745–751
5. Leff, P. (1995). The two-state model of receptor activation. *Trends in pharmacological sciences*, 16(3), 89–97.
6. Wang, A., Zhang, Z., & Zhao, Q. (2009). In vivo measurement of protein functional changes. *International journal of biological sciences*, 5(5), 411.
7. Kenakin, T. (2001). Inverse, protean, and ligand-selective agonism: matters of receptor conformation. *The FASEB Journal*, 15(3), 598–611.
8. Kobilka, B. K., & Deupi, X. (2007). Conformational complexity of G-protein-coupled receptors. *Trends in pharmacological sciences*, 28(8), 397–406.
9. Millar, R. P., & Newton, C. L. (2010). The year in G protein-coupled receptor research. *Mol. Endocrinol.* 24(1), 261–74.
10. Rajagopal, S., Rajagopal, K., & Lefkowitz, R. J. (2010). Teaching old receptors new tricks: biasing seven-transmembrane receptors. *Nature reviews Drug discovery*, 9(5), 373–386.
11. Williams, K., Dawson, V. L., Romano, C., Dichter, M. A., & Molinoff, P. B. (1990). Characterization of polyamines having agonist, antagonist, and inverse agonist effects at the polyamine recognition site of the NMDA receptor. *Neuron*, 5(2), 199–208.
12. Alewijnse, A. E., Timmerman, H., Jacobs, E. H., Smit, M. J., Roovers, E., Cotecchia, S., & Leurs, R. (2000). The Effect of Mutations in the DRY Motif on the Constitutive Activity and Structural Instability of

the Histamine H2Receptor. *Molecular Pharmacology*, 57(5), 890–898.
13. Rosen, O. M., Herrera, R., Olowe, Y., Petruzzelli, L. M., & Cobb, M. H. (1983). Phosphorylation activates the insulin receptor tyrosine protein kinase. *Proceedings of the National Academy of Sciences*, 80(11), 3237–3240.
14. Williams, D. H., O'Brien, D. P., Sandercock, A. M., & Stephens, E. (2004). Order changes within receptor systems upon ligand binding: receptor tightening/oligomerisation and the interpretation of binding parameters. *JMB*, 340(2), 373–383.
15. Chidiac, P., Hebert, T. E., Valiquette, M., Dennis, M., & Bouvier, M. (1994). Inverse agonist activity of beta-adrenergic antagonists. *Molecular Pharmacology*, 45(3), 490–499.
16. Nordlund, P., Sjöberg, B. M., & Eklund, H. (1990). Three-dimensional structure of the free radical protein of ribonucleotide reductase. *Nature* 345, 593–598
17. Zimanyi, C. M., Chen, P. Y. T., Kang, G., Funk, M. A., & Drennan, C. L. (2016). Molecular basis for allosteric specificity regulation in class Ia ribonucleotide reductase from Escherichia coli. *eLife*,*5*, e07141.

Chapter 18

The Basic Thermodynamic Principles of Enzyme Catalyzed Reaction

Summary: For simple chemical reactions, the catalyst mechanism is rather simple. However, it is more complex for an enzyme catalyzed reaction, where many chemical mechanisms are utilized at same time, which greatly reduce the activation energy of chemical reaction. This chapter will discuss the chemical mechanisms of enzymatic reaction.

Compared with non-catalyzed chemical reactions, an enzyme can accelerate the rate of chemical reaction by a factor of 10^7 to 10^{13}. Thus the puzzle about enzymology is how enzyme can obtain such great power (1). What is the catalysis mechanism of an enzymatic reaction? Here we will answer this question. The catalysis mechanisms of enzymes were extensively studied in the past and has been discussed by many textbooks in biochemistry. However, previous discussions about it are limited to the summarization of experimental results and there is no comprehensive theory (theory in thermodynamics) about it (2). In this chapter, we will discuss basic principles of it. In order to avoid tedious and lengthy explanation of trivial data and copious phenomena, our discussion is narrowed down to the basic features of several typical enzymes and the general thermodynamic principles for it.

1. The Basic Features of Acid and Base Catalyzed Reaction

At first, we examined the basic features of acid and base catalyzed chemical reactions occurring in solution. By comparing it with enzyme catalyzed reaction, the overall properties of enzymatic reaction were obtained.

Catalysis mechanism. In solution, the hydrolysis of an amide compound can be catalyzed by either acid or base, but they cannot occur at same time.

For acid catalyzed reactions, the mechanism is as shown:

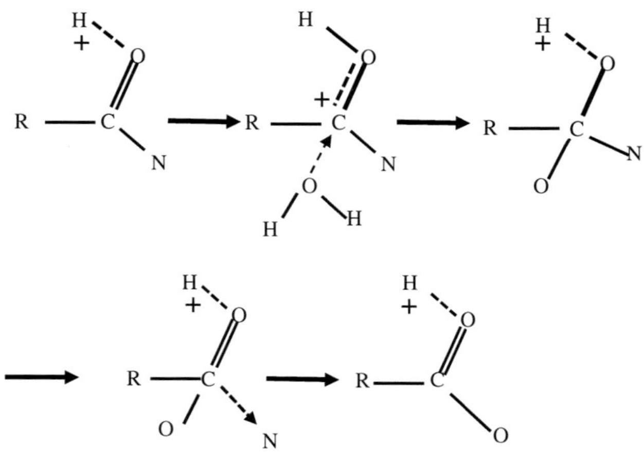

For base catalyzed mechanisms, the process of reaction can be expressed as:

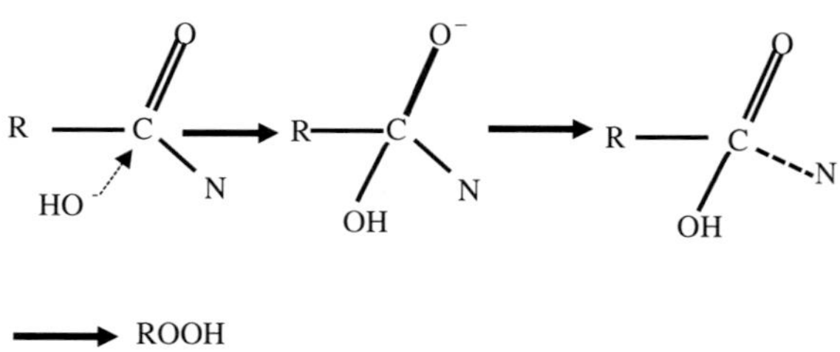

In this case, the nucleophile is hydroxyl OH^-. In contrast, in an acid catalyzed reaction, the nucleophile is the water molecule.

In solution, the breakage and formation of chemical bonds is very rapid and it is estimated to be about 10^{-12} seconds. The life of intermediate of chemical reaction is also very short. Thus the cooperation of acid and base catalysis cannot occur.

The Solvent effect: the environment of chemical reaction or the solvent of it has a great impact on the rate of reaction. It has been well studied in chemistry (3).

2. Multiple Chemical Mechanisms of Enzyme Catalyzed Reaction

Unlike simple chemical reactions, the enzyme can exploit many chemical mechanisms at the same time. The best known example is a lysozyme which utilizes both acid and base catalysis mechanism in enzymatic reactions. The Asp35 acts as an acid and the Asp52 acts as a base (4).

In chymotrypsin, the ser195 acts as the nucleophile. In chemistry, the uncharged hydroxyl is a futile nucleophile. Enzymes have developed a specific structure which is helpful to the dissociation of hydroxyl to hydroxyl and hydrogen ions. This specific structure is called the Catalytic Triad, and is formed by His57, Asp102, and Ser195 (for trypsin). The His57, as well as Ser195, are necessary for the activity of chymotrypsin, while Asp102 has the ability to stabilize the structural fashion formed between Ser195 and His57 (5, 6). In addition, the role of His57 in enzyme action can be substituted by the histidine of some substrates. It represents the conformational flexibility of the active site of the enzyme (7).

The basic mechanism of chymotrypsin is shown below (Fig. 1):

From this figure, we can see that the main mechanism of chymotrypsin was the base catalysis while the ser195 acted as the nucleophile. However, the role of several hydrogen atoms cannot be ignored. The hydrogen of ser195 and Gly103 of main back contacted with oxygen of the carbonyl group of the substrate. They may, at least theoretically, act as acids and contribute to the efficiency of enzyme action. The His57 may act as an acid and catalyze the release of product from the enzyme-reactant complex. This view was supported by the fact that mutation of Ser196Cys had very

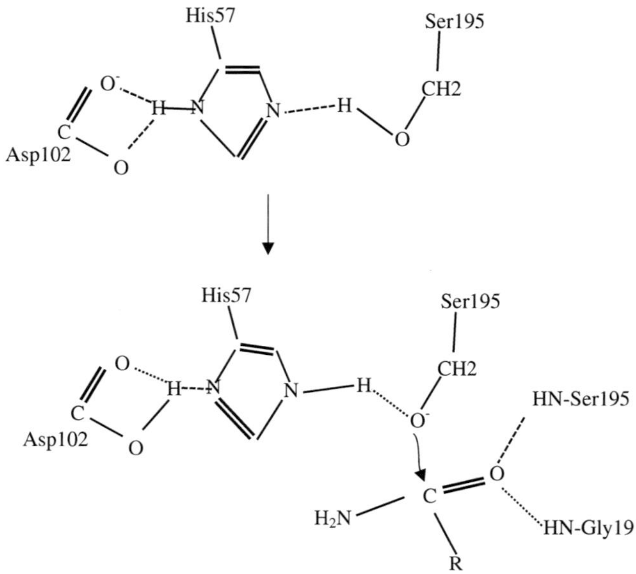

Figure 1. The diagram of working mechanism of chymotrypsin.
This Figure was drawn up from ref. 5 (with minor modification).

low activities. Thus, it is undoubtedly true that both acid and base catalysis mechanisms occur in the chymotrypsin catalyzed reaction.

Compared with that of water solution, the substrate is in a different environment when it binds to enzyme, thus the solvent effect takes a greater impact on enzymatic reaction. Professor Warshel has studied electrostatic interaction between substrate and concludes that a larger amount of energy an enzyme uses to accelerate the rate of chemical reaction comes from the electrostatic interaction between the substrate and the enzyme during the transition state of reaction (8). But it is not the sole factor that helps to accelerate the rate of chemical reaction.

3. The Energy Transferring from the Enzyme-Substrate Complex to the Transition State

Enzymes possess larger amount of energy to accelerate chemical reactions with a factor of about 10^7–10^{13}. In order to reveal the working mechanism, we must find out the source of energy to lower the activation energy of chemical reaction.

According to the dynamic model for enzyme action, many types of protein conformations are involved in enzyme catalyzed reactions (9). The role of protein conformational change involving in substrate specificity has been well documented. But the mechanism it uses to decrease activation energy has not been addressed previously.

Theoretically, when an enzyme meets a substrate in a solution, an enzyme–substrate complex is formed and it undergoes many steps of protein conformational change until transition state of chemical reaction is reached. The transition state of an enzymatic reaction is a protein conformational state of enzyme reactant complex, not the transition state of the reactant alone as that occurs in simple chemical reaction. Thus, the contribution of protein conformation to enzymatic reaction cannot be ignored.

According to the dynamic model of enzyme action, the following figures were drawn up. They represent the working mechanism for enzymatic reaction.

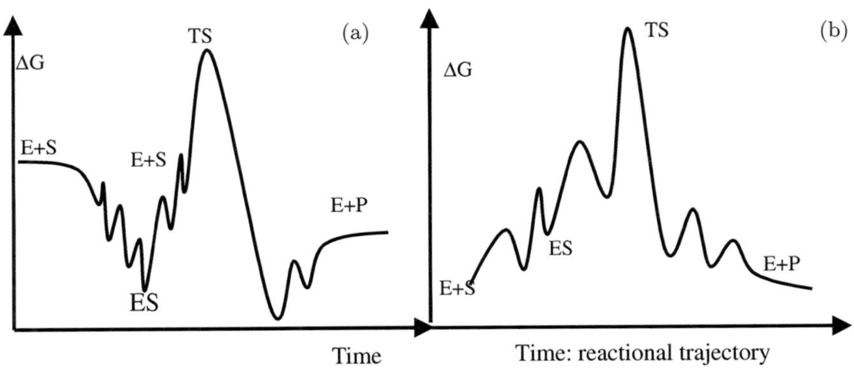

Figure 2. Diagram of the working mechanism for enzymatic reaction.

E is enzyme, S is substrate, p is product, ES is enzyme substrate complex, TS represents the transition state of reaction. Figure 2a is the energetic level of the enzyme complex. Figure 2b is the energetic level of the enzyme-reactant complex in the direction of reactional trajectory.

In this figure, we see that the energetic level of the enzyme-substrate complex is lower than that of the naked enzyme and substrate in solution (Figure 2a). However, in the aspect of reactional trajectory, the energetic level of enzyme-substrate complex is higher than that of the naked enzyme and substrate in solution (Figure 2b). In other words, there is a similarity between the protein conformation of enzyme-substrate complex and the protein conformation of an enzyme-substrate complex in transition state. In view of experimental science, the distortion of the normal conformation of a substrate in enzyme-substrate complex has been observed and it is hypothesized to contribute to the high efficiency of the enzyme. Thus, some energy of the enzyme-substrate binding has been utilized to lower the activation energy of enzymatic reaction.

Here, it is important to differentiate the potherse of enzyme-substrate interaction and the potherse related to enzyme-catalyzed-reaction or the potherse of the transitional state of enzymatic reaction. They are overlapped and differ from each other.

The energetics of enzymatic reaction can be expressed as:

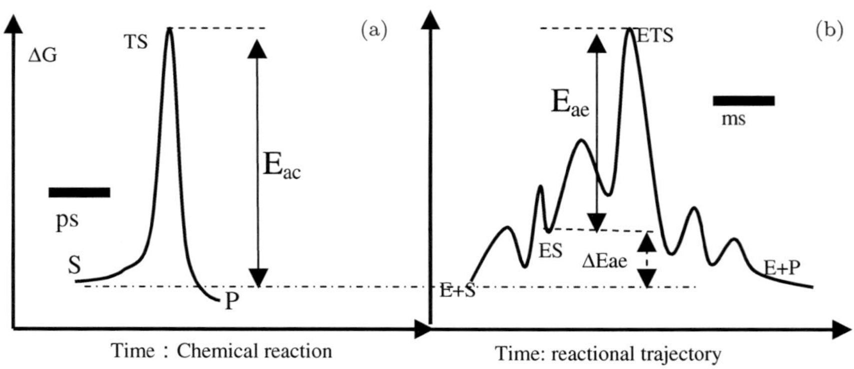

Figure 3. A diagram of energetic levels of enzymatic reactions.

E_{ac} represents activation energy of chemical reaction. E_{ae} represents the activation of enzymatic reaction. ETS represents the transitional state of the enzymatic reaction and it is a thermal state of enzyme-substrate complex. ΔE_{ae} represents the difference between activations of enzymatic reaction and chemical reaction. Figure 3a is the energetic profile of chemical reaction and Figure 3b is the energetic level of enzymatic reaction.

In Fig. 3, we see that

$$E_{ac} = E_{ae} + \Delta E_{ae}$$

$$\Delta E_{ae} \leq \Delta G_B$$

$$CP = K_D$$

Where ΔG_B is the binding affinity between enzyme and substrate. CP is the factor of enzyme to accelerate the rate of chemical reaction, K_D is the binding affinity.

For multiple working mechanisms of enzymatic reaction

$$\Delta E_{ae} \leq N(\Delta G_B)$$

$$CP = (K_D)^n$$

Where n represents the number of working mechanism involving in enzymatic reaction.

For trypsin catalyzed reaction, the K_D is about 10^4. As we have discussed, there are at least three working mechanisms for trypsin catalyzed reactions (acid: HN at backbone of Ser195; base, side hydroxyl of Ser195, deformation of substrate in enzyme substrate complex), therefore

$$CP \leq (10^4)^3 = 1^{12}$$

The mutation of Ser195Ala of trypsin decreases enzyme activity by a factor of 10^4, which is almost the same as our estimation (5).

In views of experimental science, the distortion of a substrate, the orientation of a substrate, and the proximity between different substrates have been recognized as the catalysis mechanism for enzyme action. Their contribution to enzyme action has been included in base or acid catalysis mechanisms within our model.

4. The Transition State of Enzymatic Reaction is the Product of Protein Conformational Change

A highlighted conclusion, in our theory, is that the transition state of enzymatic reaction is the transition state of enzyme-substrate complex, not the transition state of the reactant alone. Because the transition state of enzymatic reaction cannot be measured experimentally, we can only demonstrate the correctness of this conclusion by the predictions of our theory.

The transition state of enzymatic reaction is changeable. The KIE (kinetic isotope effect) is a good monitor for any changes that occurred during the transitional state of enzymatic reaction (10).

The transition state of enzymatic reaction is the product of the conformational change of enzyme-substrate. The enzyme-substrate complex shows many types of protein conformational changes and some are involved in enzyme actions while some are not logically related to enzyme actions. In order to demonstrate this hypothesis, we need to find critical evidence for it. We need to find one type of protein conformational change which is absolutely needed for an enzyme action. The mutation of trypsin has provided such an evidence. The mutation of Ser195thr of trypsin shows no activity while the mutation of C42A/C58A/S195T trypsin has high protease activity (11, 12). The underlying reason is that the s-s bond (disulfide bond between C42 and C58), which has a great impact on the pathway of protein conformational change or protein dynamics, was removed, thus the protein conformational change involved in the enzyme activity can be processed with this mutant.

5. The Hypothesis that the Enzyme Binds Tightly to the Transition State of the Reactant is Unnecessary

The so-called dogma of enzymatic reaction proposed by Pauling is that an enzyme binds tightly to the transition state of chemical reaction and thus the activation energy of enzymatic reaction was greatly reduced (13). This hypothesis was supported by facts that analogs of the substrate has a high affinity to enzyme. But that is not the critical evidence and some evidences do not support this conclusion.

According to the transitional state theory and conventional view of enzymology (1), enzymatic reactions can be represented with the following formulation:

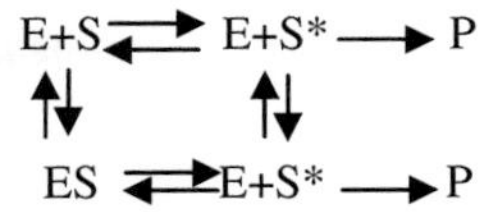

By studying the relationships of kinetic parameters in enzymatic reactions expressed in this formulation, one can demonstrate that an enzyme binds tightly to the transition state of the substrate (S*).

However, according to our model, the following formulation could be made.

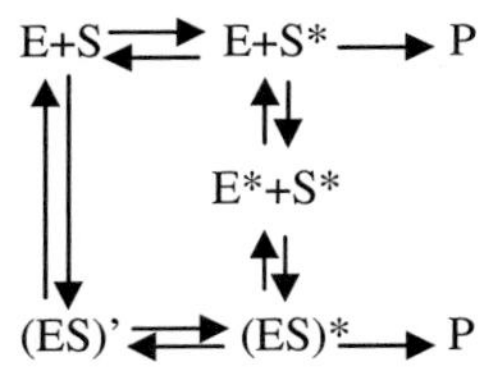

The transition conformation of enzymes (E*) differs from that of naked enzymes and enzyme-substrate complex (ES)'. Therefore, we conclude that it is unnecessary to suppose that enzymes bind tightly to the transitional state of the substrates (9).

References

1. Garcia-Viloca, M., Gao, J., Karplus, M., & Truhlar, D. G. (2004). How enzymes work: analysis by modern rate theory and computer simulations. *Science*, 303(5655), 186–195.
2. Pauling, L. (1948). Chemical Achievement and Hope for the Future. *American Scientist* 36: 50–58.
3. Welton, Thomas. Room-temperature ionic liquids. Solvents for synthesis and catalysis. *Chemical reviews* 99.8 (1999): 2071–2084.
4. Vocadlo, D. J., Davies, G. J., Laine, R., & Withers, S. G. (2001). Catalysis by hen egg-white lysozyme proceeds via a covalent intermediate. *Nature*, 412(6849), 835–838.
5. Hedstrom, L. (2002). Serine protease mechanism and specificity. *Chemical reviews*, 102(12), 4501–4524.
6. Polgar, L. (2005). The catalytic triad of serine peptidases. Cellular and Molecular Life Sciences CMLS, 62(19–20), 2161–2172.

7. Corey, David R., and Charles S. Craik. An investigation into the minimum requirements for peptide hydrolysis by mutation of the catalytic triad of trypsin. *Journal of the American Chemical Society* 114.5 (1992): 1784–1790.
8. Warshel, A., Sharma, P. K., Kato, M., Xiang, Y., Liu, H., & Olsson, M. H. (2006). Electrostatic basis for enzyme catalysis. *Chemical reviews*, 106(8), 3210–3235.
9. Zhao, Q. (2011) dynamic model of enzymes action. *Protein Pept Lett.* 18, 92–99.
10. Kohen, A., & Klinman, J. P. (1998). Enzyme catalysis: beyond classical paradigms. *Accounts of chemical research*, 31(7), 397–404.
11. Higaki, Jeffrey N., Luke B. Evnin, and Charles S. Craik. Introduction of a cysteine protease active site into trypsin. *Biochemistry* 28.24 (1989): 9256–9263.
12. Baird, Teaster T., William D. Wright, and Charles S. Craik. Conversion of trypsin to a functional threonine protease. *Protein science* 15.6 (2006): 1229–1238.
13. Fersht, A. R. (1974). Catalysis, binding and enzyme-substrate complementarity. Proceedings of the Royal Society of London B: *Biological Sciences*, 187(1089), 397–407.

Chapter 19

Protein Conformational Change and Enzyme Activity

Summary: The execution of an enzyme activity requires many steps of protein conformational change by the way of substrate binding conformation, transitional state conformation and product binding conformation. If the pathway of this protein conformational change was modified, thermodynamic and dynamic parameters of enzymatic kinetics, such as k_m, k_{cat}, activation energy, as well as the affinity between enzyme and substrate will be altered. Thus, we can judge protein conformational change related to an enzyme activity by a change of parameter of enzymatic reaction. Here the basic principles in our understanding of the behaviors of enzyme will be discussed.

In solution, a protein shows several types of conformation at global level, many at local level, and numerous at atom level; a protein wanders forever among this grant space of conformational ensemble, while in the meantime the unique structure of the protein remains unchanged (1–7). When an enzyme interacts with its substrate, a new species of protein or protein complex of enzyme-substrate is formed. This protein complex then undergoes several essential steps of protein conformational change, such as the conversion of open state and closed state of enzymes, and which will result in the formation of a protein complex of enzyme product. In this process, the transitional state conformation of an enzymatic reaction is produced and a chemical reaction is processed within it (8). Although the enzyme-substrate complex roams among its conformational ensemble without a fixed pathway of conformational change, a pathway of protein conformational change along the reactional trajectory can be theoretically isolated and any alteration of it will result in a change of enzyme activity and parameters of enzymatic kinetics. Thus, we can monitor or analyze some properties of a protein ensemble by a change of parameters of enzymatic kinetics.

This chapter will discuss the basic principles of this theory, evidence and related phenomena about it, and working mechanism for the conversion between inactive and active conformation of enzyme.

1. The Basic Principles

According to the dynamic models of enzyme action and protein thermodynamic structure theory, the following principles can be obtained, which will serve as the basic principles for our discussion of enzyme action and regulation (4, 8).

1) In solution, a protein has a complex thermal system and many types of protein conformation at global level and numerous conformations at local or atom level have been produced on the basis of numerous internal motions of its components. A protein roams within this protein conformational ensemble forever (1–6). In view of the protein thermodynamic structure theory, a subset of the protein conformational ensemble represents a thermodynamic state of a protein or a potherse (4).
2) The sample space of the conformational ensemble for any protein is very large and its properties change with protein dynamics and protein structure.
3) The enzyme-substrate binding and conversion from normal conformations of the enzyme-reactant complex to the transition state conformation of enzymatic reaction requires several steps of protein conformational changes which represents a pathway of conformational change along reactional trajectory.
4) The transitional state conformation of an enzymatic reaction also represents a subset of a protein conformation ensemble of an enzyme- reactant complex. The transitional state of an enzymatic

reaction is the transitional state conformation of enzyme-reactant, but not the transitional state of reactant as it does in simple chemical reaction.

5) All factors, inclusive of physical and chemical factors, such as gene mutation, activator, and inhibitor, take their role in enzyme action by affecting the pathway of protein conformational change for enzymatic reaction. The pathway of protein conformational change involved in enzymatic reaction may then be affected by a change of parameters of enzyme kinetics.

The relationship of diversified concepts of enzymology was expressed in Figure 1 and we can analyze the nature of protein conformational change of enzymatic reaction by applying these principles.

2. The Nature of Active and Inactive Conformations of Enzymes

In 1950s, it was observed that an enzyme can be at a state in which the enzyme is not denatured, but is not active. This state of the enzyme is termed the inactive conformation of the enzyme. Since then, the concept of active and inactive conformation of enzyme has been well accepted, but the nature of it has not been fully understood.

The dominating view about it is that there are structural differences between active and inactive conformations and it can account for the change of the enzyme activity between them. The structural difference between crystals of the inactive and active conformation of some species of enzymes have been recognized (10, 11,12). However, it is argued that such differences that have appeared in the protein crystal do not represent an actual state of a protein in solution. In addition, it is very difficult to demonstrate a logical relationship between such conformational change and enzyme activity change.

A preliminary new idea is that the impaired landscape of a protein is the reason for the inactivation of enzyme. The structural change is not absolutely required for the conversion between inactive and active conformation (13, 14, 15).

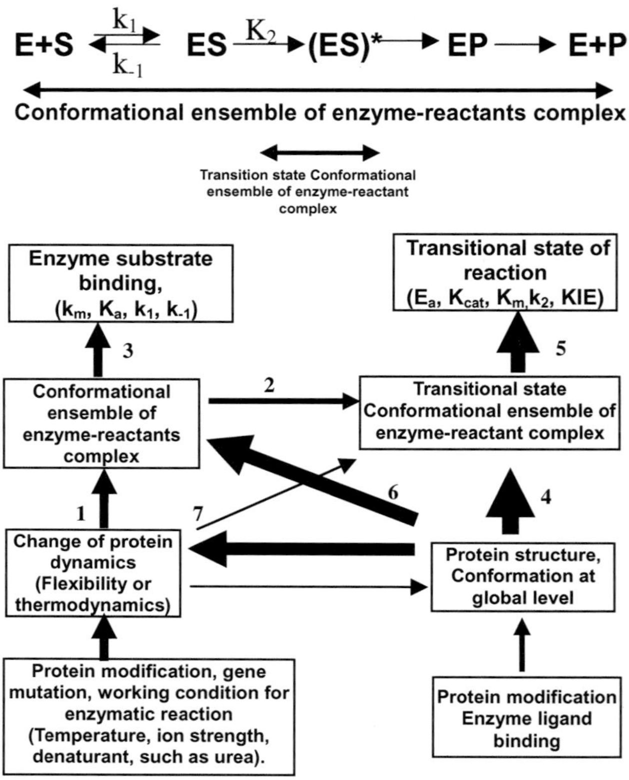

Figure 1. The logic relationship of diversified concepts of protein science and enzymology.

The process of enzymatic reaction is expressed in upper Figure. E_a is the activation energy of the enzymatic reaction, K_a is a binding constant. The thicker the arrow, the stronger the correlations between the relevant components in this Figure.

Our theory states that there is a common connection between the protein dynamics and protein conformational ensemble. The pathway of the protein conformational change of an enzymatic reaction can be influenced by a change of protein dynamics, which will result in a change of enzyme activity. The structural change is not a necessary prerequisite for conversion between active and inactive conformation, although it occurs under some cases. But the structural change of a protein has a great impact on protein dynamics, the pathway of protein conformational change, therefore a dramatic change of enzyme activity is normally accompanied with a structural change. In view of protein thermodynamics, the protein conformational ensemble represents a thermodynamic state of a protein or a potherse (4, 16).

The nature of inactive and active conformation of enzymes is shown in Figure 2. The conformational ensemble of inactive conformations of enzymes is not overlapped with the conformational

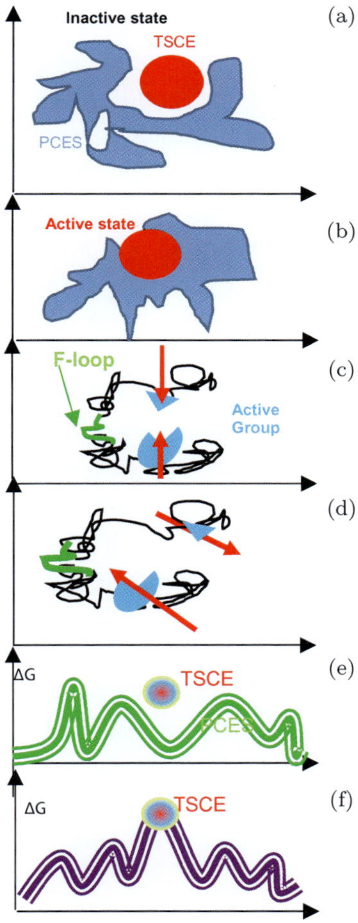

Figure 2. Canton for pathway of protein conformational change of enzymatic reaction.

ensemble of the transitional state conformation of an enzymatic reaction and thus the pathway of protein conformational change of enzymatic reaction cannot be processed.

3. The Explanations and Interpretations for the Basic Principles

The two steps involving in the pathway of protein conformational change found in an enzymatic reaction are tightly related to working mechanism of the enzymatic reaction (17–20). The first is the enzyme substrate binding, and the basic principle of it has been well studied under the name of the protein ligand interaction and we will not discuss it here (3, 4, 9). It was indicated that properties of enzyme substrate binding obtained from studying of enzymatic kinetics is same as those obtained from other studies on protein ligand interaction, such as an increase of k_m when protein flexibility is increased (10, 21–22).

The second is the conversion from normal conformation of enzyme-reactant complex to the transition state conformation of enzyme-reactant complex. The principle is shown in Figure 2.

TSCE is an inactive transition state conformational ensemble. PCES is the protein conformational ensemble of enzyme-substrate complex. Figure 2a illustrates the nature of inactive conformation. The PCES does not overlap with the TSCE. For an active conformation of enzyme, which is shown in Figure 2b, the PCSE of it is overlapped to or includes TSCE. Figure 2c and 2d express the direction of motion of protein components and its relation to the formation of TSCE. The f-loop refers to flexible loop of a protein and changing it has great impact on direction of internal motion of the protein. Figure 2e and 2f describe pathway of protein conformational change of active or inactive conformations along reactional trajectory (there is no fixed pathway of protein conformational change, so we outline it in many lines).

The TSCE represents a subset of a protein conformational ensemble. According to the protein thermodynamics structure theory, any potherse is related to a concrete property of a protein and the transitional state of enzyme reactant is a property of the enzyme, hence it represents a special potherse (4). The KIE (kinetic isotope effect of enzymatic reaction) is a sensitive monitor that can record the changing in the transitional state of enzymatic reaction. The abnormal KIE can be observed under extreme condition with the enzyme-substrate interaction still being processed well under these conditions (23–26). Another evidence comes from the studying of enzyme ligand binding; it shows that the subtle differences in enzyme structure can be further expressed in transitional state of enzymatic reaction (27). Furthermore, even the active group of enzyme can be replaced by a group of a substrate, which occurs in so-called "substrate assisted reaction" (28). These indicated that TSCE is not the fixed structure, but a thermodynamic state which permit alteration in some degree.

The Mechanism for pathway (1+3) (expressed in Figure 1). Theoretically, the sample space and diversified properties of the conformational ensemble of an enzyme-substrate complex is changeable if the protein dynamics or flexibility is altered by denaturants, such as urea. In this case, the binding affinity between enzyme and substrate is changed. As a result, the change of k_a, E_a, and km of enzymatic reaction was altered and they can be observed experimentally.

Normally, an enzyme can utilize many substrates. The conformational ensemble of an enzyme-substrate complex, as well as enzymatic reaction such as E_a, and k_m, varies from substrate to substrate.

In addition, much of the common knowledge of enzymology can be easily explained, including: a) the separation of enzyme substrate binding ability and catalysis ability; b) how the protein conformational ensemble can permit alteration in some degree without impairing its function. Thus comes the concept of structural and functional promiscuity of proteins and enzymes (21, 22).

The mechanism for pathway [4+6+2+5]: The structural change of a protein will cause a dramatic change of the protein conformational ensemble and thus has a great impact on all aspects of a protein. If it happens, the E_a and K_{cat} of enzymatic reaction would be greatly changed.

The relative probability of TSCE could be calculated. When we consider the enzymatic kinetics of an enzyme acting serial substrates, the relative probability of TSCE can be obtained (16, 29). In this case, the relative probability of TSCE is related to enzyme activity or activity.

The activation or inactivation of an enzyme can be achieved by different mechanisms. As PCES represents a system which is composed of many conformations, it can change in different ways and thus deviation can be obtained from any direction. Normally, trypsin can be activated by hydrolysis at its main chain, however, the active conformation of trypsin has been discovered in a mutant without hydrolysis at its backbone (30). Enzymes can be denatured by many factors, such as temperature, denaturants, electrostatic interaction of it with other charged substance, and the inactive conformations produced under these conditions differs from each other (31).

The disulfide bond has great impact on enzyme activity. The disulfide bond is a constraint for protein motion and it affects the capacity of the protein conformational ensemble greatly. It has been observed that the removal of the disulfide bond (cys42–58) of trypsin can endow protease activity of Ser195Thr of trypsin, which shows substrate binding ability and no protease activity (32). That is critical evidence in support of our model. Another example comes from the study of Maun (33).

The overlaps between PCES and TSCE can be judged by the parameters of enzymatic kinetics. For a substrate, if PCES and TSCE of the enzymatic reaction are overlapping well, the k_{cat} will be larger and activation energy will be small, and vice versa. It has been reported that behaviors of esterase vary when it acts as different substrates, or a favorable and an unfavorable substrate. The rate limiting step of enzymatic reaction varies with substrates (34). Our opinion is that the protein conformational ensemble for the enzyme-substrate complex and transitional state conformational ensemble of an enzymatic reaction change with substrates. The diversified nature of the enzymatic reaction for different substrates is expressed in Figure 3.

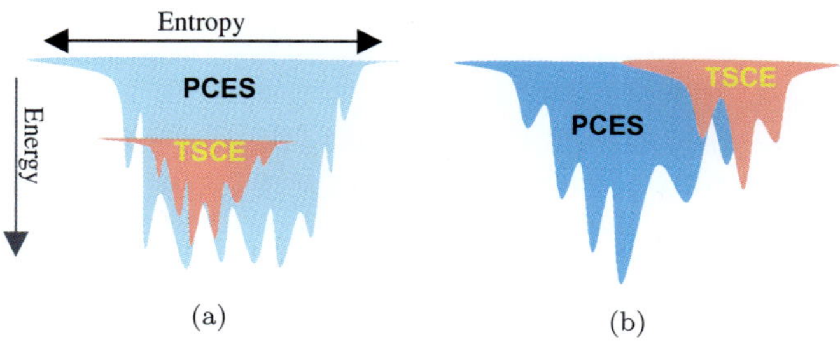

Figure 3. Landscape of PCES and TSCE in pathway of protein conformational change.

The landscape of both PCES and TSCE is substrate and enzyme (mutation) dependent. Figure 3a and 3b describe the PCES and TSCE for optimum substrate and none-optimum substrate. For the optimum substrate, the PCES and TSCE are overlapped well and conversion between them processes easily.

References

1. Frauenfelder H, Sligar S G, Wolynes P G. The energy landscapes and motions of proteins. *Science*, 1991; 254: 1598–1603.
2. Kurplus M and McCammon JA. Dynamics of proteins: elements and function. *Annu Rev Biochem.* 1983; 52: 263–300.
3. Boehr DD, Nussinov R, Wright PE. The role of dynamic conformational ensembles in biomolecular recognition. *Nat Chem Biol.* 2009; 5: 789–96.
4. Zhao Q. natures of protein thermodynamics and thermodynamics. *Rev Theor Sci.* 2013; 1: 83–101.
5. Zhao Q. Nonalgorithmicity and algorithmicity of protein science, *Adv. Biosci. Biotech.* 2011; 2(5) 340–346
6. Pitt WR, Montalvão RW, Blundell TL. Polyphony: superposition independent methods for ensemble-based drug discovery. *BMC bioinformatics.* 2014; 15: 324.
7. Kurkcuoglu Z, Bakan A, Kocaman D, Bahar I, & Doruker P. Coupling between catalytic loop motions and enzyme global dynamics. *PLoS comput boil.* 2012; 8: e1002705.
8. Zhao Q. dynamic model of enzymes action. *Protein Pept Lett.* 2011; 18: 92–99.
9. Dyer R B, Callender R. Advances in Time-Resolved Approaches to Characterize the Dynamical Nature of Enzymatic Catalysis. *Chemical Reviews*, 2006; 37: 3031–3042.

10. Somero GN. Temperature adaptation of enzymes: biological optimization through structure-function compromises. *Annu Rev Ecol Syst.* 1978; 1–29.
11. Laskowski RA, Gerick F, Thornton JM. The structural basis of allosteric regulation in proteins. *FEBS Lett.* 2009; 583: 1692–8.
12. Johnson L N, Noble M E, & Owen D J. Active and inactive protein kinases: structural basis for regulation. *Cell*, 1996; 85: 149–158.
13. Shen T, Zong C, Hamelberg D, McCammon JA, Wolynes PG. The folding energy landscape and phosphorylation: modeling the conformational switch of the NFAT regulatory domain. Faseb J. 2005; 19: 1389–95.
14. McCammon JA. Gated Diffusion-controlled Reactions. *BMC Biophys.* 2011; 4:4.doi: 10.1186/2046–1682–4–4.
15. Nolen B, Taylor S, Ghosh G. Regulation of protein kinases; controlling activity through activation segment conformation. *Mol Cell.* 2004; 15: 661–75.
16. Zhao Q. A thermodynamic and theoretical view for enzyme regulation. *Biochemistry (Mosc).* 2015; 80: 1–7.
17. Hedstrom L. Serine protease mechanism and specificity. *Chemical reviews*, 2002; 102: 4501–4524.
18. Hedstrom L, Szilagyi L, Rutter W J. Converting trypsin to chymotrypsin: *the role of surface loops. Sci.* 1992; 255: 1249–1253.
19. More N, Daniel R M, & Petach H H. The effect of low temperatures on enzyme activity. *Biochem. J*, 1995; 305: 17–20.
20. Fersht, A. (1985) Enzyme Structure and Mechanism, Freeman Press, San Francisco.
21. Khersonsky O, Tawfik DS. Enzyme promiscuity: a mechanistic and evolutionary perspective. *Annu Rev Biochem.* (2010); 79: 471–505.
22. Nobeli I, Favia A D, Thornton J M. Protein promiscuity and its implications for biotechnology. *Nature biotech.* 2009, 27(2): 157–167.
23. Schramm V L. Enzymatic transitional states and transitional state analog design, *Annu Rev Biochem*, 1998; 67: 693–720.
24. Kohen A, Cannio R, Bartolucci S & Klinman JP. Enzyme dynamics and hydrogen tunneling in a thermophilic alcohol dehydrogenase. *Nature*, 1999; 399: 496–499.
25. Giordano A, Febbraio F, Russo C, Rossi M, &Raia C A. Evidence for co-operativity in coenzyme binding to tetrameric Sulfolobus solfataricus alcohol dehydrogenase and its structural basis: fluorescence, kinetic and structural studies of the wild-type enzyme and non-co-operative N249Y mutant. *Biochem J.* 2005; 388: 657–667.
26. Klinman JP. &Kohen A. Hydrogen Tunneling Links Protein Dynamics to Enzyme Catalysis. *Annu Rev Biochem.* 2013: 82: 471–496.

27. Zoi I, Motley M W, Antoniou D, Schramm VL & Schwartz S D. Enzyme Homologues Have Distinct Reaction Paths through Their Transitional States. *J Phys Chem B.* 2015; 119: 3662–3668.
28. Dall'Acqua W, Carter P. Substrate-assisted catalysis: molecular basis and biological significance. *Protein Science*, 2000; 9: 1–9.
29. Zhao Q. Partition function of protein conformational state. *J Comput Theor Nanosci.* 2012; 9: 745–751.
30. Pasternak A, Liu X, Lin T Y, & Hedstrom L. Activating a zymogen without proteolytic processing: Mutation of Lys15 and Asn194 activates trypsinogen. *Biochemistry*, 1998; 37: 16201–16210.
31. Hong R, Fischer NO, Verma A, Goodman CM, Emrick T, Rotello VM. Control of protein structure and function through surface recognition by tailored nanoparticle scaffolds. *J Am Chem Soc.* 2004; 126: 739–743.
32. Baird TT Jr, Wright WD, Craik CS. Conversion of trypsin to a functional threonine protease. *Protein Sci.* 2006; 15: 1229–1238.
33. Maun H R, Eigenbrot C, Raab H, Arnott D, Phu L, Bullens S, & Lazarus R A. Disulfide locked variants of factor VIIa with a restricted β-strand conformation have enhanced enzymatic activity. *Protein Sci.* 2005; 14: 1171–1180.
34. De Simone G, Mandrich L, Menchise V, Giordano V, Febbraio F, Rossi M, Pedone C, Manco G. A Substrate-induced Switch in the Reaction Mechanism of a Thermophilic Esterase. *J. Biol. Chem.* 2004; 279: 6815–6823.

Chapter 20

Allodynamic Regulation Model for Enzyme Regulation — A Thermodynamic View

Summary: The allosteric regulation model is a tradition model of enzyme regulation. It states that a change of protein conformation at a site can be transmitted into remote areas of the protein because the conformation of a protein can change cooperatively, which is the nature of protein three-dimensional structure. On the basis of the protein thermodynamic structure theory, an allodynamic mechanism for enzyme regulation has been proposed and it is a thermodynamic model for enzyme regulation. It states that a change of protein thermodynamic state at one site can be transmitted into remote sites of the protein via protein thermodynamic structure reorganization. A comprehensive understanding of enzyme regulation can be obtained on the basis of this model.

The precise regulation of enzyme activity is crucial for a cell. Currently, the allosteric regulation model serves as the main theory for enzyme regulation. Two different models of allosteric regulation have been proposed: Monod–Wyman–Changeux (MWC) and Koshland–Nemethy–Filmer (KNF) models (1–3). These models differ in some aspects, but all agree that enzyme activity is regulatable. Recent revisions of these models include the conformation selection hypothesis, but it is not a mature theory for enzyme regulation and we will not discuss it. Both MWC and KNF models are experimental and the underlying mechanism of enzyme regulation is still unclear, thus many experimental results remain unexplainable (3). On the basis of protein thermodynamics and dynamics, we have proposed the allodynamic mechanism, which provides a basic understanding of the mechanism for enzyme regulation at a fundamental level (4–7).

1. Basic Principles of Allodynamic Regulation

Firstly, several principles of enzyme regulation, which can be deduced from the protein thermodynamic structure theory, are stated here.

1. In solution, any enzyme or protein exists in many conformations, most of which are inactive. The equilibrium among them is determined by the thermodynamic principle. In the simplest case or in a given condition, there is only one active conformation of the enzyme.
2. The abundance of protein conformations is changeable under different conditions. The abundance of one conformation can be calculated from data on protein conformational stability by applying the concept of a partition function of protein conformation (6).
3. The thermodynamic state of one conformation, such as the energetic level, is also changeable under different conditions (6).
4. A change in protein dynamics at one site can be transmitted with different degrees into other sites of a protein via protein thermodynamic reorganization, which may elicit a conformational change at other sites of a protein. However conformational change (shapeless) is not always observed.
5. Any factors, such as temperature, solvent, substrate, drug, positive and negative regulator (activator or inhibitor), or genetic mutant impact enzyme regulation by influencing the protein conformation stability (protein flexibility or dynamics) of enzymes, which controls the abundance of active conformations and thermodynamic state of the active conformations. Correspondingly, there are two related mechanisms for enzyme regulation by these factors. First, these factors alter the thermal state of each conformation of an enzyme. Second, as a result, the redistribution of thermodynamic state changes the abundance of active conformations and therefore catalytic activity.

6. Effects of diversified regulators on enzyme regulation could be integrated thermodynamically within an enzyme molecule. We will not discuss it here for it also obeys the basic principles of enzyme regulation.

A major difference between our model and the current dynamic contribution to enzyme regulation is that within our model a change in protein dynamics does not directly influence enzyme activity, but instead impact enzyme regulation by influencing the abundance of active conformations of the enzymes (8).

2. Distribution Curve of Active Conformations of Enzymes and Enzyme Regulation

The distribution curve of active conformations of enzymes is an important concept in our discussion of enzyme regulation because the nature of enzyme regulation can be analyzed based on the profile of the distribution curve.

Another concept of enzyme regulation is sensitivity of enzyme regulation. It can be defined as

$$\mathrm{da/ds}$$

Where a is enzyme activity and s represents the stimulus affecting enzyme activity, such as temperature or the concentration of protein denaturant

$$\text{or } \mathrm{da/dS}$$

Where S is the concentration of substrate.

The distribution curve of the active conformation of an enzyme cannot be measured experimentally now, but we can deduce it from the enzyme activity curve. If the Arrhenius plot of enzymatic reaction is linear, the probability of active conformation of enzymes will reach 100% in the linear range. If the Arrhenius plot is nonlinear, the probability of active conformation of enzymes will be changeable. According to the dynamic model for enzyme activity and partition function of protein conformation, the Arrhenius equation of an enzymatic reaction can be written as (6, 7):

$$K = APe^{-[\Delta E_0 + B_T(T-T_0)]/K_BT} = APDe^{-\Delta E_0/K_BT} \qquad (1)$$

On the basis of this equation and conclusions we have drawn as aforementioned, the active conformation distribution curve of active conformation of enzyme can be figured out.

In Figure 1, we have drawn up two different various of distribution curve of the active conformation of an enzyme. Figure 1a and Figure 1b may show same enzyme under different conditions. Within our model, the pH, temperature, salt, or other environmental factors, act as regulators of enzyme activity. For example, the trypsin is inactive in pH 3.0 (in stomach), but is active at pH 8.0 (in small intestine). This feature matches up its physiological functioning well.

In this figure, we can also see that the efficacy of the regulator is changeable under different conditions. It may act as a positive

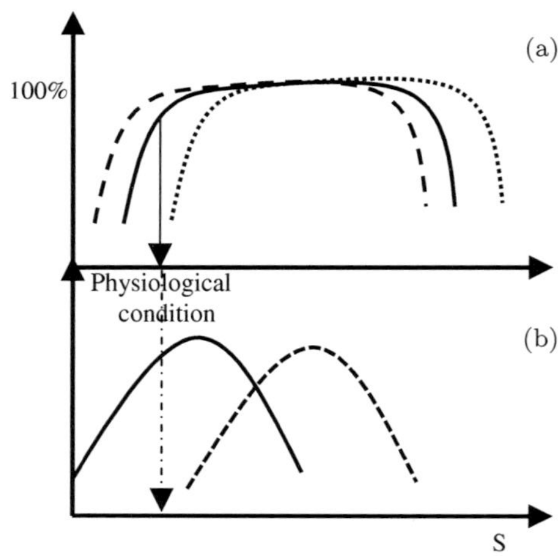

Figure 1. Distribution curve of the active conformation of an enzyme and enzyme regulation.

Where s represents value of conditional factors for enzyme activity, such as pH, temperature, and concentration of protein denaturant. Two types of distribution curves of the active conformation of an enzyme were shown in Figure 1a and 1b. When the protein conformation ensemble is influenced by protein modification or other factors, the distribution curve of active conformation can be shifted (as illustrated in dashed lines). Thus, under appropriate physiological conditions, enzyme activity is changeable and regulatable.

regulator under a condition and a negative regulator under another condition.

3. The Cooperative Binding Among Hemoglobin and Oxygen Molecules

In history, the enzyme regulation mechanism was developed based on the study of cooperative phenomenon in the binding between oxygen and hemoglobin. It has been observed that the binding curve between oxygen and hemoglobin is sigmoid. If there is no cooperation in the binding of different oxygen molecules to hemoglobin, the binding curve should be similar to that of an enzymatic reaction (plotting rate of an enzymatic reaction as function of concentration of substrate) (9). The underlying reason is that the binding of first oxygen to hemoglobin could affect the binding ability of the second oxygen molecule to hemoglobin. Or in other words, the protein conformational change at one site can be transmitted into the other sites of a protein. This mechanism is called allosteric regulation. But the fundamental thermal mechanism of it has not been revealed on the basis of allosteric regulation itself.

Our theory has revealed the thermodynamic mechanism of enzyme regulation. The theoretical foundation of our theory is not the cooperativity concept, but the protein thermodynamic structure theory. The cooperativity is a deduction concept in our theory. It is a property of potherse change. According to our model, the binding between the first molecule of oxygen and hemoglobin induces protein conformational change of hemoglobin at binding site. Some, but not all, of conformational change at the binding site can be transmitted into other areas of hemoglobin and which induces potherse change at an advanced level. This type of conformational change can be described by a concerted model (3). Another type of conformational change at the binding site of a protein, which cannot propagate into other area of the hemoglobin, cannot be described by concerted model, instead it can be illustrated using a sequential model (10).

We can deduce conclusions about enzyme regulation by applying principles of allodynamic regulation: 1) the energy difference between the T (tense) and R (relax) states is changeable in a changing condition which will result in the transition from positive and negative cooperativity and a change of Hill coefficient of enzymatic kinetics under different conditions; 2) the behaviors of cooperativity can be influenced by many factors, which in turn influence protein conformation, such as the pH of buffer (Bohr effect), protein denaturants, temperature, salt concentration etc. In fact, their effect was integrated thermodynamically and the behavior of cooperativity is the final output.

The mechanism of cooperativity can be seen clearly in Figure 2.

When the first oxygen molecule binds to hemoglobin, it induces local conformational change at the first subunit. Global protein conformational change is also induced; where by the global state of hemoglobin is changed to the relax state. The binding of other molecules of oxygen cannot induce s global conformational change of hemoglobin, but it can further induce local conformational change. When energy levels of the T and R state are converted under different conditions, the property of cooperativity (positive or negative) will be changed.

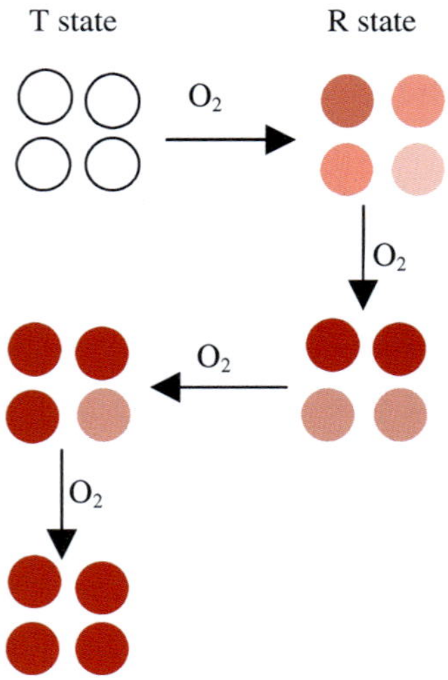

Figure 2. Diagram for allodynamic model for oxygen and hemoglobin binding. The red circle represents the R state and closed circle represents the T state of hemoglobin.

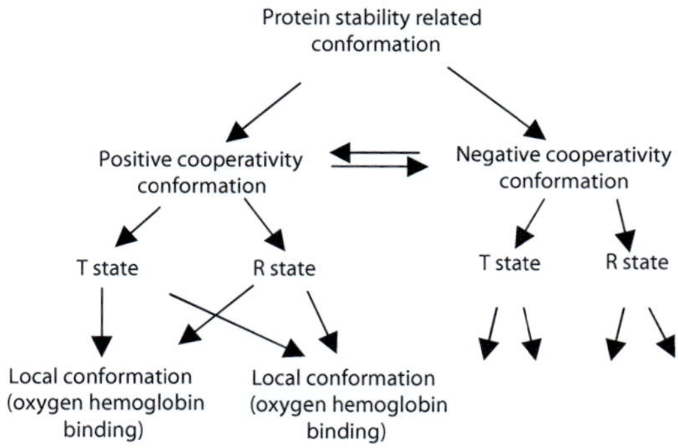

Figure 3. Hierarchical structure of protein conformation involved in the cooperativity of hemoglobin oxygen binding.

Many types of protein conformation which have been involved in the cooperativity of hemoglobin oxygen binding is shown in Figure 3. When these types of protein conformation are changed, the behaviors of protein conformation cooperation (regulation) are changed as well.

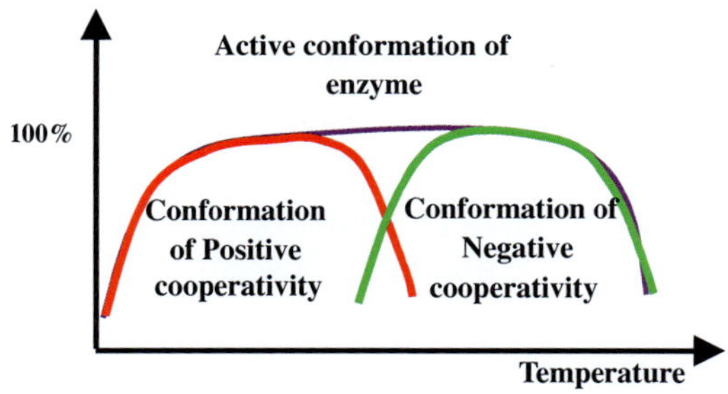

Figure 4. Transition of cooperativity of alcohol dehydrogenase.

The distribution curve of active conformation, and conformation for positive and negative cooperativity is shown in this figure.

Many types of protein conformation are involved in the execution of cooperativity of enzymatic reaction. Some of it have been pictured as follows (Fig. 3).

4. Conversion of Cooperativity Under Different Cases

Some enzymes can show both cooperativity under different condition, such as dehydrogenase (11, 12). Behaviors of the transition of cooperativity of alcohol dehydrogenase is shown in Figure 4.

The alcohol dehydrogenase shows negative cooperativity at low temperatures and positive cooperativity at high temperatures. The underlying reason is that the energy level of protein conformational states is temperature dependent.

5. Noncompetitive Inhibitor Acts Allodynamic Regulator

In view of experimental science, the noncompetitive inhibitor of an enzyme is characterized by its ability to inhibit enzyme activity while its binding ability to the enzyme is independent to enzyme substrate binding; when the noncompetitive inhibitor is in the presence of a working buffer, the k_m of enzymatic reaction remains unaltered, but the V_m is changed. In the presence of a noncompetitive inhibitor in a working buffer, the enzymatic reaction rate can be precisely described by the Michaelis-Menten equation.

The allosteric regulator of an enzyme is characterized by its ability to alter the behavior of cooperativity shown in enzymatic kinetics. Traditional view of allosteric regulation refers to oligomeric enzymes and thus the noncompetitive inhibitor (activator) is not the allosteric regulator of enzymes. In the presence of an allosteric regulator, the enzymatic reaction rate cannot be described by Michaelis-Menten equation.

In view of structural biology, allosteric regulator alters the protein structure or conformation of an enzyme at global level. Such change of protein conformation induced by noncompetitive inhibitor of enzymes has not been documented (e.g. 13). However, such a change does not meet the criteria for an allosteric regulator.

The study of the underlying mechanism of allosteric regulation results in a new definition for it: an allosteric regulator of an enzyme (or protein) binds to enzyme at site far away from the active site of the enzyme. On the basis of this definition, the noncompetitive inhibitor of an enzyme should be considered as an allosteric regulator.

These two opposite opinions were held by many scientists and student of biochemistry and we cannot reconcile the two opinions theoretically. Thus confusion was generated and there is currently still no consensus about it.

Within our model, the noncompetitive inhibitor is an allodynamic regulator (or allosteric regulator). It regulates the abundance of active conformation of enzyme in solution. For a typical noncompetitive inhibitor under a given condition, enzyme activity can be completely inhibited, but in most cases, enzyme activity can only be partially inhibited and the efficacy of the inhibitor is changeable with the changes in condition. The pH, salt, and all others factors of environment for enzyme activity can also be considered allodynamic regulator.

6. Advantages of Our Model

The analysis shows that enzyme regulation can and must be explained by the laws of thermodynamics, and our model has many advantages over the allosteric regulation mechanism. First, our model reveals a thermodynamic mechanism for enzyme regulation that cannot be provided by an allosteric model. Second, it is difficult for the allosteric model to provide a satisfactory explanation for negative cooperativity, however within our model both negative and positive regulatory behaviors of an enzyme can be shown by same enzyme under different conditions. Third, within our model, enzyme activity, enzyme regulation, and bio-information integration (cooperation or counteraction) within an enzyme molecule are processed via a common mechanism — thermodynamic reorganization of the protein — while the allosteric model cannot support such a view. Fourth, within our model all enzymes, oligomeric and monomeric, are regulatory, while an allosteric model cannot explain regulatory behaviors of monomeric enzymes. Fifth, the effect of protein dynamics on enzyme regulation is well defined in our model. Sixth, the structural change of an enzyme is not essential within our model, but it is an absolute perquisite within an allosteric regulation mechanism [7]. Seventh, our model can give a quantitative description of enzyme activity and regulation, while the allosteric model cannot. Finally, within our model, there are many allodynamic sites for an enzyme, but the allosteric site is not an essential concept for some factors, such as urea, that can exert their effect on nearly all sites of a protein.

7. Why are Enzymes so Big?

Theoretically speaking, enzyme activity is related to a small part of enzyme. A small protein, such as a lysozyme, can do all things of enzyme. However, enzymes are normally very large and some are oligomeric. Thus, the question of the size of enzymes is prevalent in biochemistry (14, 15). Apart from enzyme activity, an enzyme needs other signaling activity to function properly, for example, protein localization, but all these cannot account for the molecular weight of enzymes. So we must seek alternative explanations for it. The enzyme regulation mechanism has provided a reasonable answer for it.

Firstly, big enzymes are capital to becoming regulatable enzyme. By gene mutation at dynamic site of protein, a new ligand binding site can be formed and the enzyme can be regulated by many factors. For example, both myoglobin and hemoglobin have the ability of binding oxygen, but binding ability between hemoglobin and oxygen is regulatable. Compared to myoglobin, hemoglobin has large surface and can bind more substance. The principle of it can be seen in Figure 5.

Secondly, compared to small enzymes, big enzymes are sensitive to physical factors of macro-world, such as mechanical force.

Thirdly, the regulation of big enzymes can be processed stably and smoothly. For example, big enzymes have more charge than small

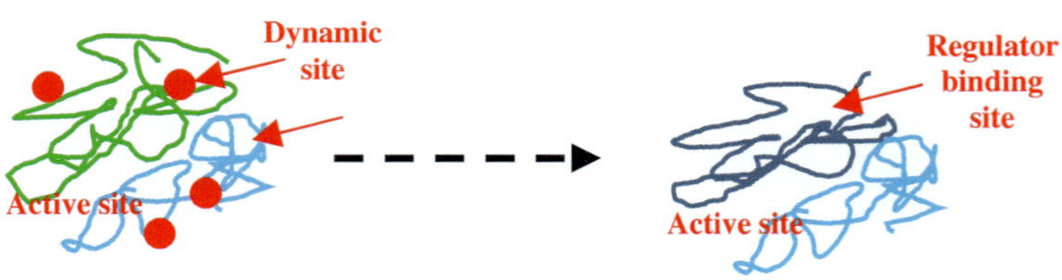

Figure 5. The generation of regulatable site of enzyme.

Dynamic sites of an enzyme have a great impact on enzyme activity. When it was changed into a regulator binding site by gene mutation, the enzyme became regulatable. Big enzyme can become regulatable enzyme without changing other properties of it. Small enzyme can also do it, but other properties of enzyme will be changed too.

proteins, and the loss of one charge (by a change of pH or protein modification) have little to do with property of big enzymes. But it may fundamentally change properties of a small protein.

Fourthly, a big enzyme binds tightly to substrate and which is helpful toward enzyme activity. Big enzymes can undergo long-range conformational change while small enzymes cannot achieve.

Fifthly, for oligomeric enzymes, enzyme activity can be enriched in a point of space. So it may facilitate the regulation of signaling activities for a cell. In this case, the enzyme itself acts as a scaffold protein.

Due to these reasons, big enzymes have many types of advantage and are hence more often selected during biological evolution.

References

1. Koshland Jr, D. E., Nemethy, G., & Filmer, D. (1966). Comparison of experimental binding data and theoretical models in proteins containing subunits. *Biochemistry*, 5(1), 365–385.
2. Monod, J., Wyman, J., & Changeux, J. P. (1965). On the nature of allosteric transitions: a plausible model. *Journal of molecular biology*, 12(1), 88–118.
3. Changeux, J. P. (2012). Allostery and the Monod-Wyman-Changeux model after 50 years. *Annual review of biophysics*, 41, 103–133.
4. Zhao, Q. (2009) Protein thermodynamic structure. *IUBMB life*, 61, 600–606
5. Zhao, Q. (2013) Allodynamic Regulation of Protein Activity. *Quantum Matter*, 2, 144–152
6. Zhao, Q. (2012) Partition function of protein conformational state. *J. Comput. Theor. Nanosci.* 9, 745–751
7. Zhao, Q. (2015) A thermodynamic and theoretical view for enzyme regulation, *biochemistry (Moscow)*, 80, 1–7.
8. Kern, D., & Zuiderweg, E. R. (2003). The role of dynamics in allosteric regulation. *Current opinion in structural biology*, 13(6), 748–757.
9. Hill, A. V. (1910). The possible effects of the aggregation of the molecules of haemoglobin on its dissociation curves. *J Physiol (Lond)*, 40, 4–7.
10. Koshland, D. E. (1995). The key-lock theory and the induced fit theory. *Angewandte Chemie International Edition in English*, 33(23–24), 2375–2378.

11. Koshland, D. E., & Hamadani, K. (2002). Proteomics and models for enzyme cooperativity. *Journal of Biological Chemistry*, 277(49), 46841–46844.
12. Giordano, A., Febbraio, F., Russo, C., Rossi, M., & Raia, C. X. (2005). Evidence for co-operativity in coenzyme binding to tetrameric Sulfolobus solfataricus alcohol dehydrogenase and its structural basis: fluorescence, kinetic and structural studies of the wild-type enzyme and non-co-operative N249Y mutant. *Biochem. J*, 388, 657–667.
13. Prade, L., Engh, R. A., Girod, A., Kinzel, V., Huber, R., & Bossemeyer, D. (1997). Staurosporine-induced conformational changes of cAMP-dependent protein kinase catalytic subunit explain inhibitory potential. *Structure*, 5(12), 1627–1637.
14. Srere, P. A. (1984). Why are enzymes so big? *Trends in Biochemical Sciences*, 9(9), 387–390.
15. Britt, B. M. (1997). For enzymes, bigger is better. *Biophysical chemistry*, 69(1), 63–70.

Chapter 21

The Protein Evolution

Summary: A protein can change its properties and transform itself into another species of protein during biological evolution. This chapter discusses the basic and general principle of protein evolution.

The study of protein evolution is of great significance both in technology and basic science (1, 2). In technology, obtaining the desired properties of a protein is the long term goal of protein engineering. Here we will discuss some problems of it by applying the principles of protein thermodynamic structure theory.

In view of genetics, protein evolution is processed by a change in protein sequence, including amino acid residue substitution, insertion or deletion of polypeptide within a protein (1). The gradual change in protein sequence leads to the drift of protein structure and shape, which finally results in the generation of new protein with new biological functions and structures.

1. The Basic Principles

We have learnt several principles from the protein thermodynamic structure theory about protein science and the following are the basic principles of protein evolution.

First, the structure of a protein shows its dynamic nature and a protein shows many types of conformation in solution. Some types of protein conformations are in low abundance in solution and their functioning have not been exploited by organism. But these conformations may become fully functioning during biological evolution.

Second, anyone property of a protein, such as protein stability and function, is only logically related to a small number of residues of the protein. Thus the change of protein sequence may change a property of the protein without influence on other properties of a protein.

Third, any two properties or conformations of a protein can be integrated or co-structured within a protein. For example, the immunological protein IgG can link all types of antigen as a form of immune response. Accompanied with this nature, the IgG is a flexible protein. Its variable area can undergo subtle structural changes or residues substitution which can made it capable to bind many different antigens (3). In addition, the IgG is big protein and the change in variable or flexible area cannot destroy its whole structure. The structural flexibility of IgG has been well documented in the textbook of immunology.

2. The Origin of the Protein

Theoretically, there are two different ways for the generation of a new species of a protein. The first way refers to the generation of new species of protein which is created from unfolded polypeptides without any biological function, such as unstructured polypeptides. But at present we know nothing about it. In biology, the generation of immune proteins, such as IgG, may be generated by such mechanism.

The second way is via the modification of old species of proteins. From where a protein is originated, it can undergo structural drift, functional drift, and transform into other species of protein with different biological functions, such as enzyme activity (4, 5). The mechanism of it is shown in Figure 1.

3. The Evolution of a Property of a Protein

When a microorganism is placed in a new environment, its proteins will evolve so that the properties of new proteins can adapt to the living environment. In other words, new proteins will show its optimum activity under living condition (6, 7). Microorganisms which can grow and survive under extreme environments are called

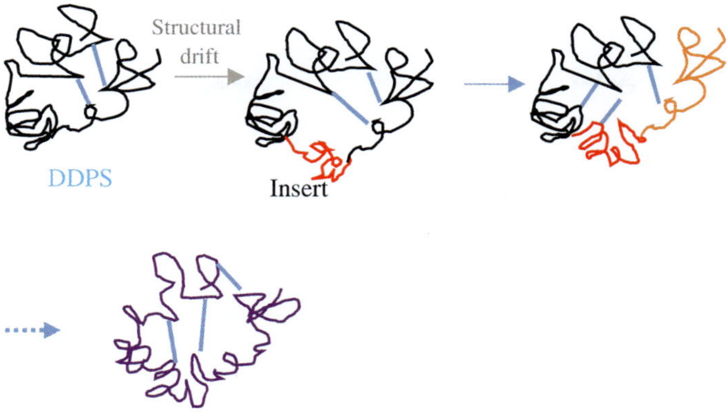

Figure 1. The diagram of protein evolution.

DDPS represents the dominated determinants (interaction) of a protein structure and it controls the fashion of the unique protein structure. First, a protein sequence is subjected to change (insertion in the figure) and it results in a structural drift or conformational drift. It then undergoes further change and structural transformation, where a new type of protein structure is generated. The structure or biological function of a protein can be further refined by gene mutation.

extremophile. The best environmental factors to be studied are temperature, salt, pH, oxygen concentration, and pressure. Normally, enzymes isolated from thermophiles are very stable and have higher activity rate at high temperatures near its living temperature. In contrast, enzymes isolated from mesophiles are less stable and have higher activity rate at low temperatures. The concept of the distribution curve of active conformation of enzymes can account for this phenomenon. According to this model, the stability of enzymes of thermophile is high, thus the active conformation of the enzyme remains unchanged at high temperatures. In contrast, the active conformation of enzymes of mesophile will become very low at high temperature. The distribution curves of active conformation of enzymes has been drawn in Figure 2.

In this figure, we can see that the living temperature of the microorganism is in a range where the active conformation of enzyme is near 100%. In order to obtain high activity from the enzyme, the optimum temperature of an organism is nearly at higher

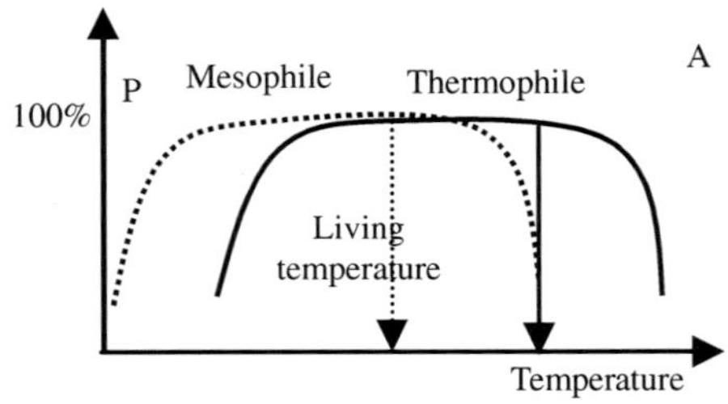

Figure 2. The diagram of the shift of the distribution curve of active conformation of enzymes among mesophile and thermophile.

P is the probability of active conformation of enzymes.

temperatures, at which protein irreversible denaturation does not occur; or in other words, the protein stability of the protein is marginal in living condition. This feature is most obvious for signaling proteins or enzymes. Compared with enzymes of mesophiles, the distribution curve of active conformation of enzymes of thermophiles have been shifted into high temperatures (8).

There is a dispute about the mechanism of protein adaptation to environmental change. One view is that it is biological rule for the enzymes of thermophiles to show low activity at low temperatures. Another view is that this phenomenon is merely a reflection of a physical rule, in other words, enzymes, which can show high stability, will show low enzyme activity at low temperature due to low protein flexibility at low temperatures, which is required for enzyme activity.

Our view is that both views are reasonable to some degree. For example, the alcohol dehydrogenase from a thermophile begins cold denaturation at 30°C, so its activity at low temperatures is due to loss of protein stability of active conformation (9). Of course, it is possible to design more stable enzymes which are active at low temperatures, but their sequences and structures will differ from that of the native protein in principle. In fact, this type of enzyme has been observed in nature (10).

Overall, the conformational shift of proteins and enzymes is a common phenomenon and it can be clearly seen in many cases. It is the basic principle for protein adaptation to a changed environment.

Table 1. The substrate specificity of ADH isolated from different species (data from Ref. 9, 10).

	ADH1	ADH2
Substrate	relative activity	
Ethanol	100%	100%
1-Propanol	82%	100%
1-Butanol	50%	93%
1-Pentanol	85%	74%
2-butanol	120%	8.3%
2-Pantanol	112%	5%
Allyl alcohol	102%	128%
Benzyl alcohol	40%	6.5%

The enzyme activity is defined 100% when ethanol is used as substrate.

4. Protein Structural and Functional Promiscuity and Evolution

Protein functional promiscuity can be clearly seen when we compare the substrate specificity of enzymes (11, 12). Enzymes such as alcohol dehydrogenase can utilize many different substrates, but the parameters of enzymatic reaction differ from substrate to substrate. This reflects the structural and functional promiscuity of enzymes.

In Table 1, we have summarized the relevant activity of two species of alcohol dehydrogenase for different substrates.

In this table, we see that alcohol dehydrogenase isolated from different organisms have different substrate spectrums although they share the same activity on a substrate. It reflects the subtle difference among their structure, conformation, and protein dynamics.

When we considered the protein functional and structural promiscuity, a possible pathway of protein evolution can be obtained and it is shown in Figure 3. This mechanism for protein evolution takes advantage of the fact that a protein can remain its function in process of biological evolution.

One important evidence for this mechanism is that some species of alcohol dehydrogenase show weaker activity of acetaldehyde

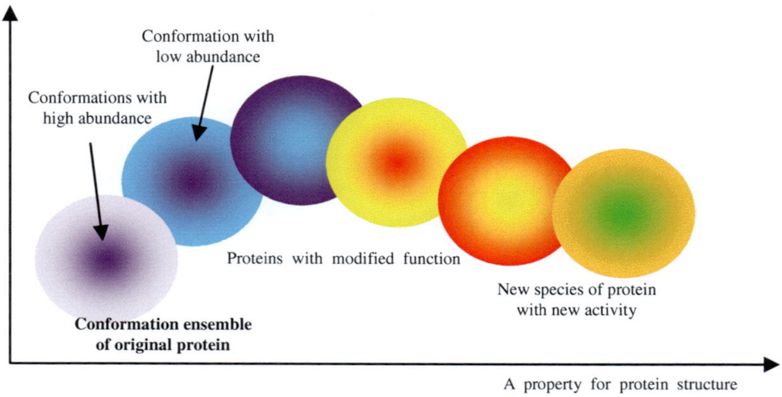

Figure 3. The protein conformation shift and protein evolution.

Each protein shows many types of conformation. When a protein is subjected to mutation, and its conformational ensemble is changed, the conformation of low abundance may become of high abundance for the mutated protein. This will induce a change of substrate spectrum of enzymes. When more mutation occur, new enzyme may be produced.

dehydrogenase (12, 13). For most species of alcohol dehydrogenase, the activity of acetaldehyde dehydrogenase is very low. When a gene mutation of alcohol dehydrogenase occurs, it may show low activity of acetaldehyde dehydrogenase. Then it may rapidly evolve into acetaldehyde dehydrogenase with full activity. Of course, the alcohol dehydrogenase may be generated from acetaldehyde dehydrogenase during biological evolution.

5. Protein Stability and Evolvability

The protein stability is a thermodynamic measurement for protein structural stability or protein flexibility. If the stability of a protein is high, it will be harder to deform and its evolutional rate will be low (14, 15). This feature can be clearly seen by comparing distribution curves of active conformation of mutated protein and wild protein. The feature of protein structural stability is shown Figure 4.

In Figure 4a, we can see that changes of the abundance of active conformation or activity for a given gene mutation of an enzyme with

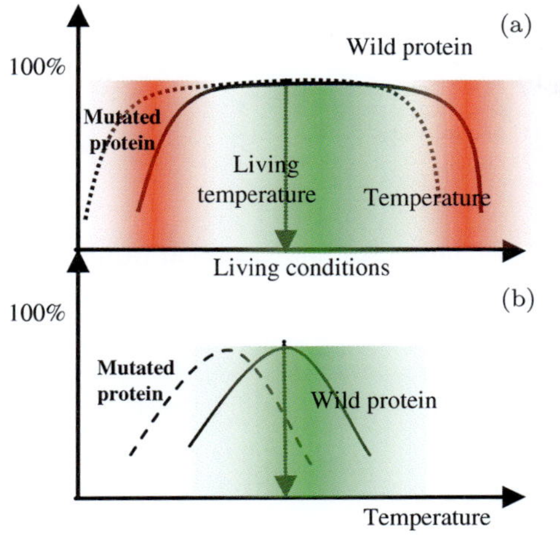

Figure 4. Protein stability and protein evolution.

The figures show two distribution curves of active conformation of a protein and a mutated protein. The protein stability is high in Figure 4a. The differences of activities between two proteins occurs in red range, which is far away from the living condition. In contrast, the activity difference occurs in living conditions if the protein has low stability (Figure 4b).

high stability occur under extreme conditions that are far away from the living condition of an organism. In contrast, changes of enzyme activity or abundance of active conformation of enzymes with low stability occur under the living conditions of organism (Figure 4b). Thus, protein with low stability is prone to be mutation. This trait can be selected during biological evolution.

6. Coupling between Two Properties (or Signal) of a Protein

The generation of regulatable proteins and enzymes is an important process of biological evolution. Theoretically, protein regulation is related to two properties of a signaling protein. The first is the regulator binding conformations of the signaling protein. The second is the coupling between the binding conformation and the effector conformation. In order to produce an efficient regulatable protein or enzyme, these two conformations should be structured

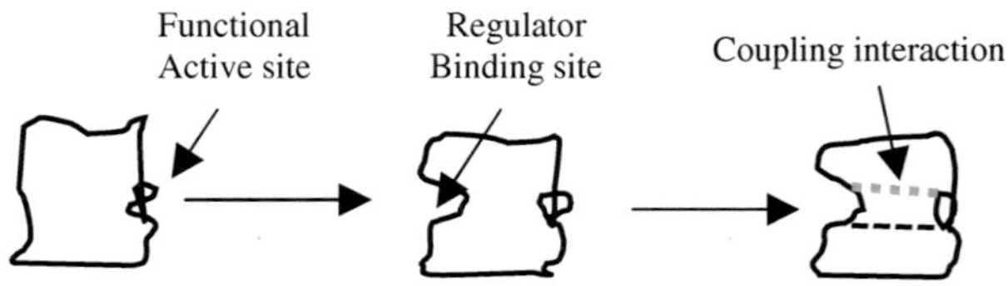

Figure 5. The generation of protein regulation fashion.

If the primordial of a protein has no regulation fashion, it can be generated in the process of biological evolution. This occurs when, firstly, a regulator binding site of the protein was produced by way of genetic mutation and this protein can further transform into regulatable protein by introducing coupling interaction between two sites.

well and coupled with each other dynamically. With that, a change of one conformation at one site can induce the change of another conformation at other sites.

According to the protein thermodynamic structure theory, the local conformation of a protein has independence to some degree without influencing conformations at other sites. Thus, by way of gene mutations, the regulator binding conformation can be generated at one site of a protein without influencing the whole structure of the protein or the conformation at other sites of the protein. The process of the generation of protein coupling or regulation fashion is shown in Figure 5.

The organisms have explored this principle in biological evolution. The immune proteins, IgG and IgM, have variable area which can bind antigen (3). All types of antigens can be linked up to the immune response by this way.

Theoretically, all protein and enzyme activity can be regulated. But only a small number of proteins or enzymes are regulatable. My opinion is that the protein structure and properties are selected during protein evolution and only key proteins that are at the intersections of signaling pathways are selected to be regulating proteins. It is a biological rule. Therefore, physical rules of protein evolution cannot give a complete explanation for protein evolution.

References

1. Grantham, R. (1974). Amino acid difference formula to help explain protein evolution. *Science, 185*(4154), 862–864.
2. Lutz, S. (2010). Beyond directed evolution—semi-rational protein engineering and design. *Current opinion in biotechnology, 21*(6), 734–743.
3. Janeway, C. A., Travers, P., Walport, M. J., & Shlomchik, M. J. (2001). *Immunobiology: the immune system in health and disease* (Vol. 2). London: Churchill Livingstone.
4. Krishna, S. S., & Grishin, N. V. (2005). Structural drift: a possible path to protein fold change. *Bioinformatics, 21*(8), 1308–1310.
5. Turner, N. J. (2009). Directed evolution drives the next generation of biocatalysts. *Nature chemical biology, 5*(8), 567–573.
6. Rothschild, L. J., & Mancinelli, R. L. (2001). Life in extreme environments. *Nature, 409*, 1092–1101.
7. Campanaro, S., Treu, L., & Valle, G. (2008). Protein evolution in deep sea bacteria: an analysis of amino acids substitution rates. *BMC evolutionary biology, 8*(1), 313.
8. Danson, M. J., & Hough, D. W. (1998). Structure, function and stability of enzymes from the Archaea. *Trends in microbiology, 6*(8), 307–314.
9. Guagliardi, A., Martino, M., Iaccarino, I., De Rosa, M., Rossi, M., & Bartolucci, S. (1996). Purification and characterization of the alcohol dehydrogenase from a novel strain of Bacillus stearothermophilus growing at 70 C. *The international journal of biochemistry & cell biology, 28*(2), 239–246.
10. Kazuoka, T., Oikawa, T., Muraoka, I., Kuroda, S. I., & Soda, K. (2007). A cold-active and thermostable alcohol dehydrogenase of a psychrotorelant from Antarctic seawater, Flavobacterium frigidimaris KUC-1. *Extremophiles, 11*(2), 257–267.
11. Tawfik, O. K. A. D. S. (2010). Enzyme promiscuity: a mechanistic and evolutionary perspective. *Annual review of biochemistry, 79*, 471–505.
12. Pietruszko, R., Crawford, K., & Lester, D. (1973). Comparison of substrate specificity of alcohol dehydrogenases from human liver, horse liver, and yeast towards saturated and 2-enoic alcohols and aldehydes. *Archives of biochemistry and biophysics, 159*(1), 50–60.
13. Heinstra, P. W. H., Eisses, K. T., Schoonen, W. G. E. J., Aben, W., De Winter, A. J., Van der Horst, D. J., . . . & Thörig, G. E. W. (1983). A dual function of alcohol dehydrogenase in Drosophila. Genetica, 60(2), 129–137.

14. Bloom, J. D., Labthavikul, S. T., Otey, C. R., & Arnold, F. H. (2006). Protein stability promotes evolvability. *Proceedings of the National Academy of Sciences*, *103*(15), 5869–5874.
15. Lachowiec, J., Lemus, T., Borenstein, E., & Queitsch, C. (2015). Hsp90 promotes kinase evolution. *Molecular biology and evolution*, *32*(1), 91–99.

Chapter 22

A Thermodynamic Model for Bio-signals

Summary: The scientific definition of molecular bio-signals has been made on the basis of protein thermodynamics in this chapter and properties of protein bio-signals will be discussed as well. Within this model, any type of bio-signal corresponds to one potherse or a thermal system within a signaling protein. The signaling activity of a protein represents its physiological activities at a fundamental level. Any physiological activity at different levels of biology is composed of many types signaling activities of proteins.

All physiological activities are processed thermodynamically. From the perspective of the signal theory, all physiological changes of an organism result of signaling activities of cellular components. The signaling activity at molecular level represents the physiological activity of organism at fundamental level. Proteins carry out all biological functions *in vivo* and the thermodynamic change of a protein is called potherse change. Thus, on the stand of protein thermodynamics, any molecular signaling activity corresponds to one thermodynamic state of a signaling protein or a potherse of it (1–3). Therefore, behaviors of cellular signaling can be described via properties of signaling proteins. This is the key principle of our theory.

From the perspective of empirical science, scientists have defined signal as all factors that can stimulate a change in the physiological state of living creatures, such as temperature, drug, ligand or agonist of receptor (enzyme), as well as the regulator of proteins.

On the standing of protein science, they all can induce the protein conformational change of a protein and trigger signaling activity of a protein *in vivo*. We agree very much with this view. Most importantly, a comprehensive understanding of signal theory on the standing of protein thermodynamics can be obtained and the logic gap between protein thermodynamic and physiology can be bridged.

1. The Definition of Molecular Bio-signal

Firstly, we must consider some basic facts or views in biology that have been well established in history:

(1) Any cellular signaling activity (or physiological system) represents a thermodynamic system.
(2) All biological functions are carried out by proteins or a protein complex.
(3) Potherse, or protein conformational state, represents the thermal state of a protein and the fashion change of protein thermodynamics are carried out by the potherse change of a protein. In molecular biology, a receptor activation results from the protein conformational transformation from the inactive state to active state. The cellular signaling activity is triggered by protein conformational transformation from inactive state to active state.

The basic principle of signal theory, or the definition of biosignal can be obtained on the basis of these recognitions.

Any signaling activity corresponds to one potherse of a protein or a thermodynamic state of a protein. The change of the cellular signaling pathways represent a change of thermodynamics and it is incurred by the potherse change of signaling protein.

Thus, there is a logical correspondence between the protein thermodynamic state and the physiological state of a living creature.

According to this definition, the signaling transduction activity can be recognized as the protein conformational coupling among different protein species, while the signal condition is protein

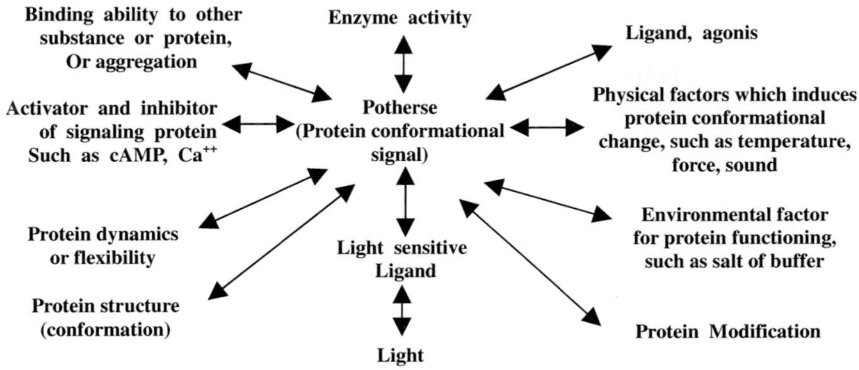

Figure 1. The Experimental and theoretical definition of bio-signal.

The central term of this Figure is the theoretical definition of bio-signal. Outside terms are experimental definitions of bio-signal.

conformational coupling among different molecules of a protein species. The protein conformational coupling can be carried out by many mechanisms: (1) direct protein coupling, which is well known in field of biochemistry; (2) by the common ligand, such as ca^{++}; (3) by common physical factors, such as voltage.

In this definition, the signals can be further recognized as any factor, which can influence the protein conformational change, such as the ligand of a receptor or regulator of a protein. In fact, this experimental definition of molecular signal is widely accepted by experimental scientists. In order to give a clear explanation, the logical relationship between the potherse and its associated properties is shown in Figure 1.

In this Figure, we can find that the experimental definition of bio-signal is understood as the behavior or properties of potherse. Although the behaviors and properties of diversified types of experimental signals vary from one to one, they all can induce the protein conformational change of a signaling protein.

2. The Properties of Bio-signals

The properties of a signal can be analyzed by this definition and the principles of protein thermodynamic structure theory. The signaling activity of a protein is triggered by a protein conformational change and it is called molecular switch.

The four typical fashions of a molecular or signaling switch, which have been extensively studied by empirical researches, are shown in Figure 2. But the general mechanism of these fashion of a molecular switch is the same. The general properties of bio-signals can be obtained by analyzing the distribution curve of active conformation (or signaling conformation) and comparing it with other properties of protein conformations.

The activation of the receptor induced by the agonist has been well documented. Its principles are shown in Figure 2a. The binding of the ligand to a receptor induces the protein conformational change and the active conformation of the receptor is produced. The receptor will be either in inactive or active conformation and there are no intermediate states between them. In other words, the transformation from inactive to active conformation is processed concertedly.

The activation of the voltage-sensitive channel represents another type of molecular switch. The signaling activity of protein flexibility or protein dynamics can be clearly seen in Figure 2b. Theoretically, the protein flexibility cannot act as the signal alone, but it can modulate the sensitivity or stability of a bio-signal by influencing the rate of the protein conformational change and the rheobase for channel opening.

Figure 2c and 2d describe the general profile of molecular and signaling switch in response to the temperature and urea stimulus. The role of the protein flexibility in signaling activity has been fully expressed in these Figures. The protein flexibility is connected with the stability of protein conformation and the change in protein flexibility can alter the stability of many protein conformations and thus influences the signaling activity of a protein.

The signaling activity of a protein can be analyzed using many properties or concepts. But the transition from inactive to active conformation of a protein is crucial for signaling activity as the signaling activity is triggered in this step. The logic relationship of many concepts in signaling activation is shown in Figure 3.

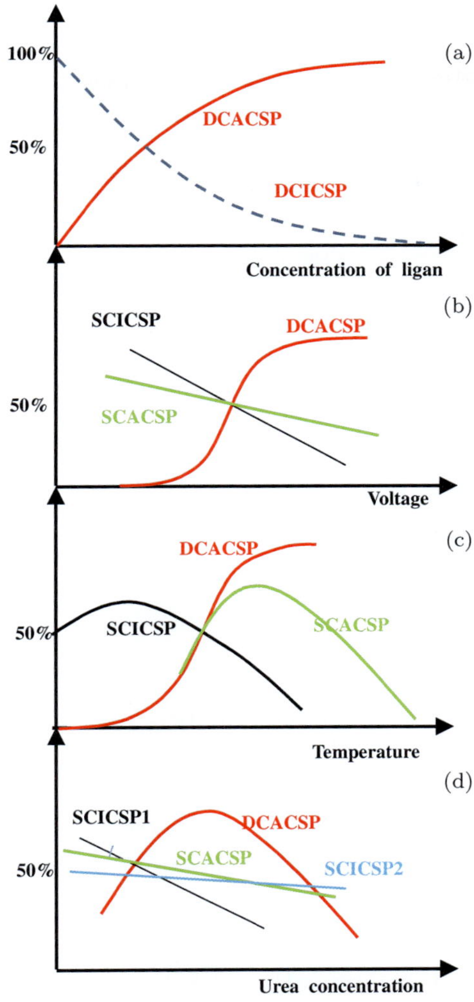

Figure 2. Four typical fashions of signaling switch.

The DCACSP is the distribution curve of the active conformation of the signaling protein, DCICSP is the distribution curve of the inactive conformation of the signaling protein, SCICSP is the stability curve of the inactive conformation of the signaling protein, SCACSP is the stability curve of the active conformation of signaling protein. Figure 2a represents the ligand triggering switch which occurs for the receptor and all proteins. The ligand binds to its receptor and the receptor is then transformed into the active conformation. Figure 2b represents properties of a voltage sensitive channel. Figure 2c outlines the general profile of a temperature sensitive receptor or protein. Figure 2d describes the general profile of a receptor in urea solution (or denaturant).

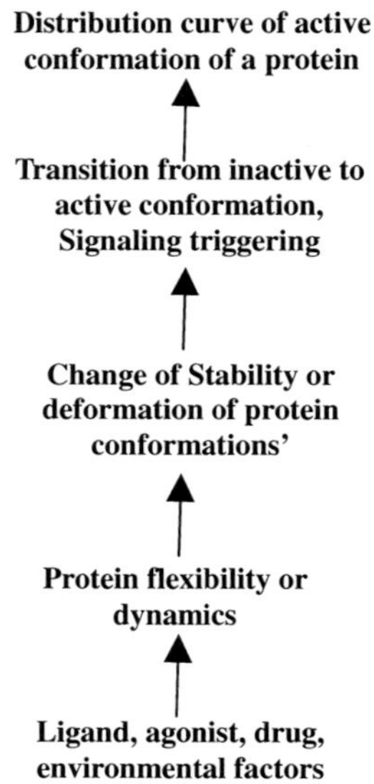

Figure 3. The logic relationship of the diversified concepts of signaling activity of a protein.

As we can see in Figure 3, the protein flexibility or dynamics (internal motion of a protein, mobility of a protein) is fundamental for any signaling activity. If the change of protein dynamic is small and signaling activity cannot be triggered, it can regulate the sensitivity of signaling activity to another type of stimulus such as an agonist; if the change of protein dynamic is larger, it can trigger the signaling activity of a protein alone.

3. The Protein Conformational Coupling in Signaling Transduction

As our definition of signal reflects a departure from current views about it, we give an explanation of signal transduction so that the reader can understand the meanings of the signal concept. The general profile of the protein conformational coupling in signal transduction pathway is shown in Figure 4.

In the pathway of signal transduction, the property of the signal can be changed and thus all types of physiological activities of the organism can be linked. For example, the agonist is a chemical

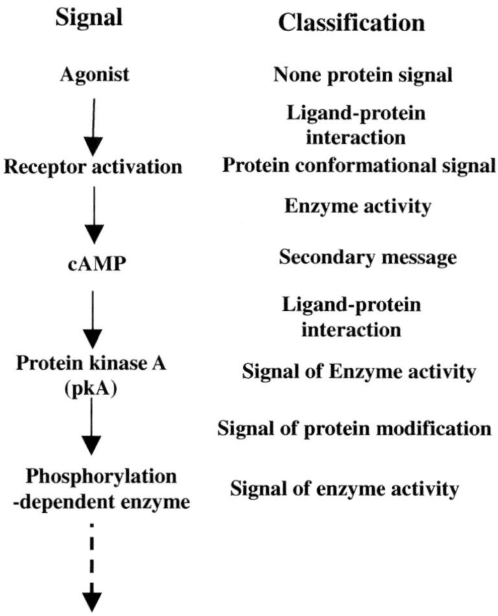

Figure 4. The pathway of signal transduction.

This Figure outlines the general profile of signaling trasduction pathway. The enzyme activity represents a special protein conformational signal.

substance and it acts as the ligand of the receptor; within the ligand-receptor complex, the chemical signal (or agonist) is transformed into the enzyme activity of the receptor which represents one type of biosignal.

4. The Classification of Bio-signal

According to the properties of bio-signals, we can classify diversified types of signals as follows:

1. Specific signal: it refers to the agonists or substances which can induce a full response from a signaling protein, such as a receptor. Normally, this type of signal acts as a specific protein at the specific site of it.
2. Nonspecific signal: it refers to the substances which cannot induce a fully response from a signaling protein in most cases, it can

regulate the sensitivity of a signaling protein. Normally, it can act as many species of proteins. Examples of this group are inhibitors, activators of enzyme or receptor.

3. Hot signal: it refers to a substance or a physical variable which can increase protein flexibility, such as temperature, protein denaturants and organic substances with tiny molecular weight (volatile anesthetics).
4. Cold signal: it refers to a substances or a physical variable which can decrease protein flexibility, such as the organic substances with high molecular weight, cold temperature and pressure.

5. The Integration of Many Types of Signal with Signaling Protein

A signaling protein can respond to many types of bio-signals at the same time. When it does, the information of the different types of bio-signals will be integrated thermodynamically within the signaling protein and a final signal can be produced.

Theoretically, the signal integration can be fully explained by the concept of the active conformational stability of the signaling protein and the distribution curve of it. At present, the relevant quantitative analysis is still insufficient. But we can judge it using different typical phenomena observed in the field.

1. The cooperation among different signals (4). If two types of signals share similar effects on protein conformational change, their effects on the signaling activity of a protein can be cooperated. Theoretically, the effects of an activator and agonist of a receptor are cooperatively on the receptor activation; they all increase the probability of the active conformation of a signaling protein. Another example is that the effects of different volatile anesthetics can be added up during the anesthesia induction of organism (5).
2. The counteraction among different signals. If different types of signals have the opposite effect on the protein conformational change of a signaling activity, their effects on a signaling protein can be countered by each other; in other words, the effect of one signal can be neutralized or suppressed by the effect of another

signal. For example, pressure can reverse the effect of volatile anesthetics (5).

3. The change in sensitivity of a signaling protein to a stimulus or agonist. The nature of signal integration within a signaling protein is changeable and is environmentally dependent. If a type of signal, such as temperature, act as an environmental factor, it can alter all the properties of a signaling protein, such as the sensitivity of signaling protein to ligand or the potency of the agonist. For example, at low concentration, urea can increase the activity of a GABA receptor, but at high concentration, urea decreases its activity.

6. The pH as a Bio-signal

The activity of cellular signal transductions is crucial for a cell and this topic of science has attracted much attention so far. Many types of pathways of signal transduction have been studied and we will not discuss it here (6–8). Hereafter in this chapter, we will discuss some types of signals which have not been well studied by other scientists.

The acid concentration of the solution (pH) can affect the functioning of all proteins. Hence it is an important signal for biological regulation. However, the pH of body fluid is under tight regulation and normally fluctuates from 7.35–7.45 for the human body. But the pH value of the environment of a living creature is changeable. Some microorganisms can survive under extreme pH condition. Thus its proteins must have evolved and can function well in the extreme conditions.

The pH of gastric acid is about 1.5–2.0, and trypsin, a digestive enzyme, is very stable under such a condition but has no activity. Under conditions of the intestinum tenue (pH 8.0), trypsin show high activity. In contrast, the pepsin is active under the condition of the existence of gastric acid. In this case, the pH acts as an environmental factor for enzyme activities.

The pH sensitive signaling activities has been isolated, but we need more evidence to assess its significance in cellular signaling.

Table 1. Difference between specific and nonspecific role of salt on signaling activity of a protein.

	Specific signal	nonspecific signal
Working concentration	0–100 mM	0–5M
Role	ligand	environmental factor
Principles	chemical reaction	partition function
Replaced by similar ion	No	Yes
Additive property	No	Yes
Cooperativity	No	Yes

7. The Salt as a Bio-signal

The salt of fluids is essential for protein signaling activity (9). Salt can act as bio-signals via two ways or mechanisms. Within the first mechanism, the salt can act as a ligand of the signaling protein and can either activate or inhibit signaling activity. Within the second mechanism, it can act as an environmental factor of a signaling protein and its effect can be described by the partition function of the protein conformational state (or nonspecific signal). Differences between the two mechanisms are expressed in Table 1.

If the salt has a nonspecific role in the signaling activity of a protein, its effect can be replaced by a similar ion. For example, the effect of Na^+ can be replaced by K^+, NH_4^+; Cl^- can be replaced by Br^-, I^-.

The general profile of enzyme activity (signaling activity) in salt solution is shown in Figure 5. In this case, the salt acts as nonspecific signal.

8. The Electric Field Acts as a Bio-signal

The protein is a charged material and there is electric field around a protein. In addition, the ion diffusion can produce electric fields within a living body. So, the electric field exists everywhere within a cell. There is no reason to doubt the importance of it in the biological functioning of the body. In fact, life has explored much of its properties in biological regulation. The electric field is essential for the proper functioning of many types of signaling proteins

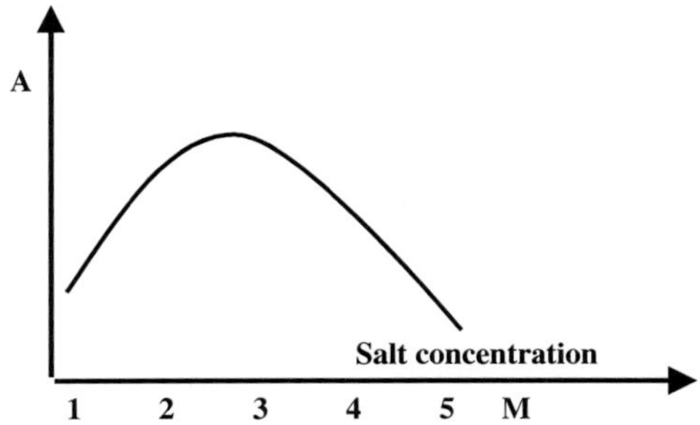

Figure 5. The general profile of the signaling activity in the presence of salt. A is the signaling activity of a protein.

(receptors). In mitochondria, the mitochondrial membrane potential (MMP) controls the protein conformation of membrane proteins (10). If MMP is changed, apoptosis of the cell may result (11).

In neural cells, the electric filed controls conformation of many channel and receptors (12).

Theoretically, the electric force can be applied to all proteins in the electric field. Under an electric field, a receptor or channel can undergo a protein conformational change, rather than a polarization.

9. The Hydrophobic Signal

The different components of a protein are arranged via hydrophobic interaction among them. Evidence has indicated that protein stability mainly comes from the hydrophobic interaction among its constituents (13). Normally, a protein has a hydrophobic core while hydrophilic residues are located on the surface of it. When a protein is in an unfavorable environment, its structure will be changed and more hydrophobic residues will appear on the surface of it. Thus, the hydrophobic surface is the signal for the protein thermal state and conformational state (14, 15). In fact, the life has exploited this property of a protein and several special signaling transduction pathways have evolved (15, 16).

The hydrophobic surface of a protein is a special signal for living condition of a cell. Some chaperons bind to hydrophobic surface of a protein and the inhibit protein denaturation (17, 18). The exposition of the hydrophobic surface of a protein, at least, take a role in the heat stress of a cell.

The protoreaction of protoplasm is another type of signaling activity induced by the exposition of the hydrophobic surface to water. In short, when a cell is stressed, the hydrophobic surface of a proteins is increased and these proteins are prone to be aggregated with each other (19).

References

1. Zhao, Q. (2010). Protein flexibility as bio-signal. *Crit. Rev. Eukaryot. Gene. Expr.* 20, 71–84.
2. Zhao, Q. (2013). A molecular and biophysical model of the bio-signal. *Quantum Matter.* 2, 9–16.
3. Zhao, Q. (2013). Nature of protein dynamics and thermodynamics. *Rev. Theor. Sci.* 1, 83–101.
4. Massagué, J. (2000). How cells read TGF-β signals. *Nature reviews Molecular cell biology*, 1(3), 169–178.
5. Evers, A. S. and Crowder, C. M. (2006). Cellular and molecular mechanisms of anesthesia. In: Barash PG., Cullen BF, Stoelting RK. ed. Clinical Anesthesia, 5th ed. Philadelphia: Lippincott Williams & Wilkins, pp. 111–132.
6. Balgi, A. D., Diering, G. H., Donohue, E., Lam, K. K., Fonseca, B. D., Zimmerman, C., ... and Roberge, M. (2011). Regulation of mTORC1 signaling by pH. *PLoS One*, 6(6), e21549–e21549.
7. Cohen, L., Asraf, H., Sekler, I. and Hershfinkel, M. (2012). Extracellular pH regulates zinc signaling via an Asp residue of the zinc-sensing receptor (ZnR/GPR39). *Journal of Biological Chemistry*, 287(40), 33339–33350.
8. Arnett, T. R. (2008). Extracellular pH regulates bone cell function. *The Journal of nutrition*, 138(2), 415S–418S.
9. Xiong, L., Schumaker, K. S. and Zhu, J. K. (2002). Cell signaling during cold, drought, and salt stress. *The plant cell*, 14(suppl 1), S165–S183.
10. Chen, L. B. (1988). Mitochondrial membrane potential in living cells. *Annual review of cell biology*, 4(1), 155–181.
11. Ly, J. D., Grubb, D. R. and Lawen, A. (2003). The mitochondrial membrane potential ($\Delta\psi$m) in apoptosis; an update. *Apoptosis*, 8(2), 115–128.

12. Long, S. B., Campbell, E. B. and MacKinnon, R. (2005). Voltage sensor of Kv1. 2: structural basis of electromechanical coupling. *Science*, 309(5736), 903–908.
13. Munson, M., Balasubramanian, S., Fleming, K. G., Nagi, A. D., O'Brien, R., Sturtevant, J. M. and Regan, L. (1996). What makes a protein a protein? Hydrophobic core designs that specify stability and structural properties. *Protein science*, 5(8), 1584.
14. LaPorte, D. C., Wierman, B. M. and Storm, D. R. (1980). Calcium-induced exposure of a hydrophobic surface on calmodulin. *Biochemistry*, 19(16), 3814–3819.
15. Otto, K. and Silhavy, T. J. (2002). Surface sensing and adhesion of Escherichia coli controlled by the Cpx-signaling pathway. *Proceedings of the National Academy of Sciences*, 99(4), 2287–2292.
16. Wahlberg, J. M. and Spiess, M. (1997). Multiple determinants direct the orientation of signal–anchor proteins: The topogenic role of the hydrophobic signal domain. *The Journal of cell biology*, 137(3), 555–562.
17. Das, K. P. and Surewicz, W. K. (1995). Temperature-induced exposure of hydrophobic surfaces and its effect on the chaperone activity of α-crystallin. *FEBS letters*, 369(2), 321–325.
18. Bukau, B. and Horwich, A. L. (1998). The Hsp70 and Hsp60 chaperone machines. *Cell*, 92(3), 351–366.
19. Matveev, V. V. (2005). Protoreaction of protoplasm. *Cell Mol Biol*, 51(8), 715–23.

Chapter 23

Protein Flexibility as a Biosignal

Summary: Protein dynamics or flexibility plays an important role in all types of signaling activities of proteins for it alters protein conformational stability and the rate of protein conformational change. In this chapter, we will discuss some signaling activities or biological functions of protein dynamics. The experimental results have indicated that a change of protein flexibility induced by volatile anesthetics is the basis of the working mechanism for anesthesia induction.

Protein dynamics or flexibility plays an important role in the signaling activities of all proteins and is involved in all types of singling activity of an organism for it controls the sensitivity of a signaling protein to a stimulus and rate of protein conformational change. Thus, we need to study the role of protein flexibility in signaling activity in detail.

Firstly, it must be pointed out that protein flexibility and protein dynamics are not physical variable and can describe different properties of a protein under different cases or conditions. Although concepts such as the distribution curve of active conformation of a signaling protein and protein stability can account for all signaling properties of a protein, there is little data now about it. In practice, we need to analyze general profile of the protein stability and distribution curve of active conformation of signaling protein by concept of protein flexibility. The protein flexibility can act biosignal by three basic mechanisms. Firstly, if a change of protein dynamics is larger enough, it can trigger the signaling activity of a protein alone, such as the activation of a receptor. Secondly, if signaling protein is at its active conformation, the change of protein flexibility can alter the quantitative properties of signaling activity for it controls the rate of protein conformational change; one example is that enzymatic activity is greatly enhanced at high temperature. Thirdly, it can alter sensitivity of signaling protein to a specific stimulus, such as agonist. So, the concept of protein flexibility is important in our analysis of protein conformational change and signaling properties of a protein, particular in field of biochemistry and molecular biology.

There are numerous signaling pathways in a cell and they have different sensitivity to a change of protein dynamics. For a given biological function or phenomena, such as anesthesia, only the property of a signaling protein which shows highest sensitivity to the anesthetics, can be expressed in the change of biological function or behavior of an organism. Other signaling activities, although they work well *in vivo*, can be ignored in the induction of anesthesia.

1. Protein Flexibility and Stability of Signaling Activity

A change of protein dynamics can be induced by many physical variables and chemical substances, such as temperature, protein denaturants (1–3). Three typical factors, temperature, protein denaturant (such as urea), and small organic substance (volatile anesthetics), have been documented and their features will be briefly discussed here. Although all these factors or substances increase protein flexibility, their underlying mechanisms differ from each other. The cooperativity among these factors can be seen in some cases, such as during protein denaturation, conformational change rate. Here we will briefly outline consensus and differences among effects of them on protein flexibility.

Firstly, temperature can directly alter the energy level of the internal motions of a protein (or protein dynamics and flexibility). In contrast, protein denaturants, as well as small organic substance, can indirectly alter protein dynamics at some sites of a protein, as they can change protein conformation.

Secondly, protein denaturants can alter protein volume to a high degree for they can bind to a protein at many sites (4); in contrast, temperature has little impact on protein volume.

Thirdly, the product of a protein at high temperature (or when there exists a high concentration of small organic substance) is protein aggregation; in contrast, the product of a protein in a high concentration of urea is extended polypeptide which shows little or no structure.

Fourthly, the exposition of the hydrophobic surface of a protein can be induced by all of these factors.

Fifth, urea can destroy hydrophobic interaction while temperature has little impact on hydrophobic interaction.

The logic relationship between protein flexibility and function has been well studied in previous chapters. Briefly speaking, protein dynamics can alter the stability and distribution curve of protein conformations and can then influence protein function indirectly. A simple explanation is shown in Figure 1.

In Figure 1, we can see that relationship between protein flexibility (dynamics) and protein function is complex. Under some cases (within PSA), a change of protein flexibility is favorable to the functioning of a protein; under other cases (within NSA), enhancing protein flexibility impairs the functioning of a protein; within the IA area, the change of protein flexibility is unrelated to the functioning of a protein.

2. Protein Flexibility and Exposition of Hydrophobic Surface

Normally, a globular protein has a hydrophobic core, within which many side chains of hydrophobic residues interacted each other via hydrophobic forces; in contrast, the hydrophilic residues are on the surface of a protein. When a protein is under an unfavorable condition, the structure of a protein can be deformed and hydrophobic surface of the protein may be produced. A protein with hydrophobic surface is prone to aggregation. Protein aggregation is believed to the cause of some diseases (5, 6). The exposition of hydrophobic

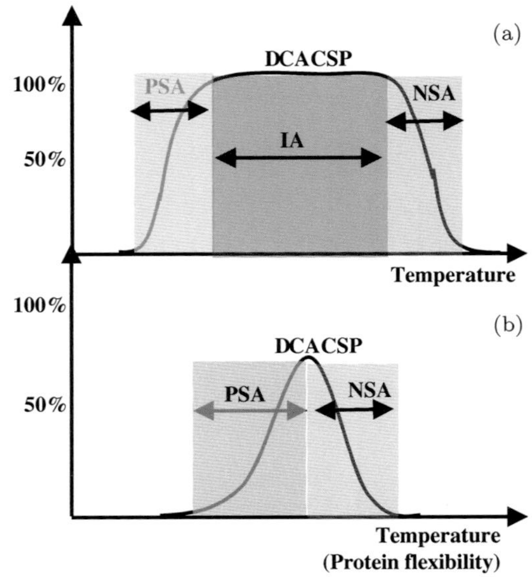

Figure 1. General profile of protein flexibility and function.

DCACSP is the distribution curve of the active conformation of a signaling protein, which is proportional to activity of a receptor (but not for enzyme activity). PSA is positive sensitive area within it a change of protein flexibility (dynamics) enhances the activity of a protein. NSA is a negative sensitive area within which a change of protein flexibility reduces the activity of a protein. Within the area IA, protein function is independent of protein dynamics or flexibility. Figure 1a is a profile of the activity curve or distribution curve of a signaling protein which has high stability. Figure 1b is a profile of the activity curve or distribution curve of a signaling protein which has low stability.

residues is also the underlying mechanism for the protoreaction of protoplasm (7).

The relationship between protein flexibility and the exposition of hydrophobic surface has been expressed in Figure 2.

3. Regulatable Protein has High Protein Flexibility or Low Stability

A cell or an organism contain numerous species of proteins; some of which are very stable, some of them are less stable.

Generally speaking, constituent proteins, such as constituent enzymes, are very stable. For example, the lysozyme of human has

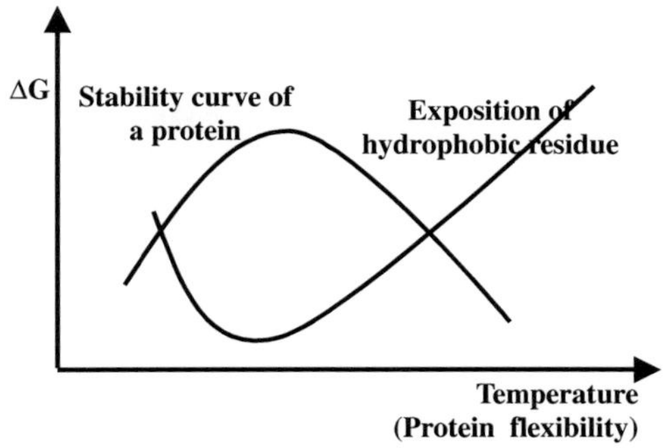

Figure 2. Logical relation between protein flexibility and exposition of hydrophobic residue.

four disulfide bonds which greatly enhances the thermal stability of lysozyme. But the stability of regulatable proteins or enzymes are very low. For example, the thermal stability of p53 is rather low (8). The relationship between protein stability (flexibility) and biological function has been well documented.

The stability of the active conformation of regulatable proteins and enzymes is low, so functioning of these proteins is sensitive to a change of environment. As a result, all distribution curves of the active conformation of regulatable protein can be classified into type 2 as pictured in Figure 3. This Figure is common for receptor activity.

As we can see in Figure 3, a change of protein flexibility can affect signaling activity of a protein by two different ways. The first mechanism is shown in Figure 3a. In a changing environment, such as temperature, protein flexibility can modulate the sensitivity of a signaling protein to a stimulus. The second mechanism is shown in Figure 3b; the distribution curve of the active conformation of a signaling protein and its mutant are located at different positions along a variable (which alters the protein dynamics of a protein) and the distribution curve of the mutant can be recognized as the curve of original signaling protein, but shifted into the other position; as a result, the activity of signaling protein will dramatically change at a given condition (test condition in Figure 3b). Many mutants of receptors with constitutive activity at physiological conditions have

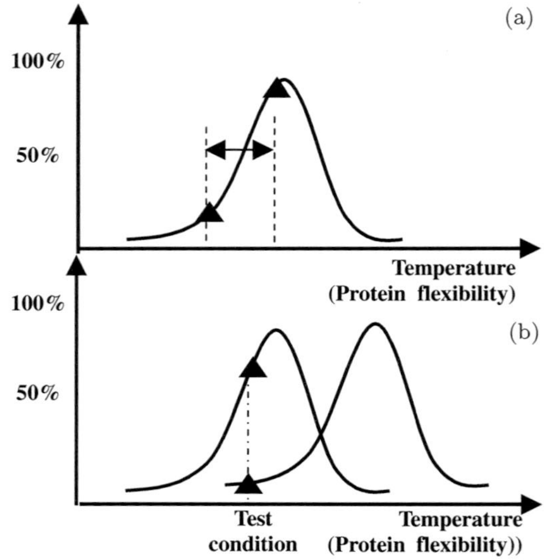

Figure 3. Distribution curves of the active conformations of signaling proteins and protein flexibility.

The general profile of distribution curves of the active conformations of signaling protein was drawn in Figure 3a. The shift of distribution curve of the active conformations of the signaling protein (for gene mutant or protein modification) along a variable (which is induced by a change of protein dynamics) was drawn in Figure 3b.

been isolated; within our model, the protein flexibility of receptor was influenced by gene mutation and they can show high activity at physiological condition (9).

4. Urea has Multiple Effects in Vivo

In field of biochemistry, urea has been recognized as a powerful material which can enhance protein flexibility. It has been confirmed experimentally that fish can endure high concentration of urea and thus the effect of protein flexibility of a signaling protein can be judged by measuring a change of fish behaviors (10). When the fish was cultured in water containing high concentration urea (0–1.2M), changes of many types of behaviors of fish can be observed. These include the loss of sensitivity to stimulus, such as loss of sound and

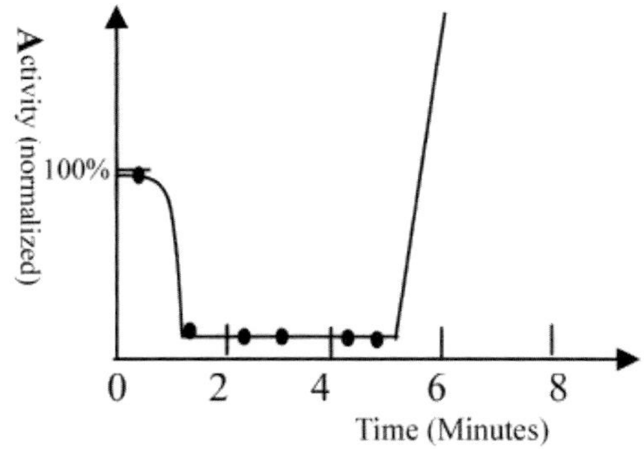

Figure 4. Sedative effect of urea.

A represents the activity of fish. The activity of fish was measured by recording the frequency of fish movement; 100% represents fish activity in the absence of drugs (40 turns/minute, waving tail). The solution contained 60 mg/L of ketamine and 0.6 M urea. The fish used were Beijing summer grass carp. The movements of fish were counted and the frequencies of tail waving were calculated. Ketamine had no impact on movement of fish; therefore, urea was seen to have a strong sedation potency. The fish became very excited after 5 minutes and counting became impossible. The curve was drawn based upon experimental data as well as behavioral observation (simulation). (cited from Ref. 10)

mechanical stimulus, sedation. The urea has strong effect of sedative effect which is shown in Figure 4.

When the fishes were put into water containing ketamine and urea. Sedative effect was first shown in fish behaviors. After a longer duration, the fish became excited as it induced a signaling activity.

5. Protein Flexibility and Anesthesia Induction

When the fish was put into the urea solution, its protein flexibility was enhanced greatly. Multiple behavior changes of fish were observed. But some types of physiological change cannot be full expressed externally or behaviorally. In order to measure such change, we need to elicit it behaviorally by adding some drug which can strengthen the effect of urea on a given behavior of the fish (10).

For example, the anesthesia state of the fish cannot be observed even high concentration of urea was added into water. But the dose of

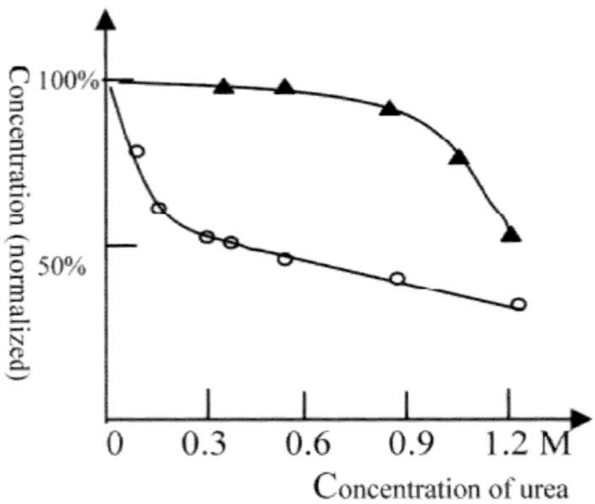

Figure 5. Quantity of anesthetics required for anesthesia production in the presence of urea.

The normalized concentration of ketamine (○) or ethanol (▲) required for anesthesia production in the fish in the presence of urea. The concentration of ketamine (125 mg/L) or ethanol (2% by volume) required for anesthesia production in the absence of urea was 100% (in order to analyze and compare results between different tests, we explored normalized concentration, rather than absolute concentration). The temperature was 22°C and Beijing summer grass carp were used in this test.

anesthetics has been greatly reduced for the production of anesthesia in urea solution.

In Figure 5, we can see that the cooperation between protein flexibility and ketamineon anesthesia induction occurs in range of urea concentration from 0–0.3 M; in contrast, the cooperation between urea and ethanol occurs in urea concentration from 0.9–1.2 M. This result indicated that the NMDA receptor (mina target protein of ketamine) is not the main target of volatile anesthetics. This result also shows that urea has weak potency of anesthetics.

6. Temperature and Protein Flexibility

Temperature is an important physical variable for an organism. The change of temperature has great impact on the functioning of all proteins. But different species of protein have different sensitivity at a given temperature; the sensitivity of proteins differs largely at different temperatures (10).

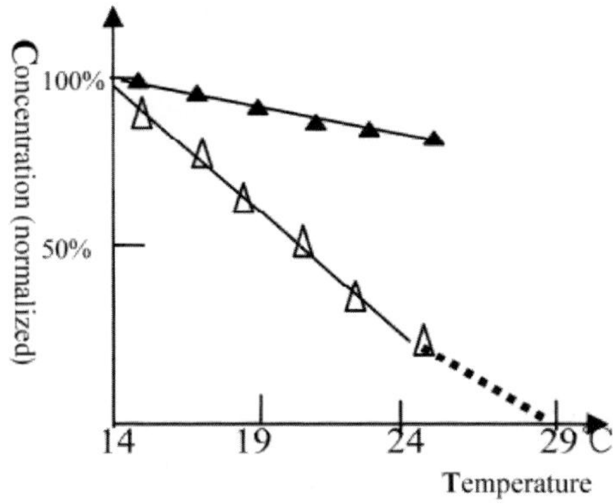

Figure 6. The dose of ketamine or ethanol at different temperature to produce the anesthesia ate of fish.

(△) represents ketamine and (▲) represents ethanol. The concentration of ketamine (150 mg/L) or ethanol (2.6% by volume) required for anesthesia production at 14° was 100%.

The cooperation between temperature and anesthetics on the induction of anesthesia of the fish is shown in Figure 6.

At 27 or 29°C, the fish become very excited and the measurement of anesthesia became impossible. The ketamine concentration required for induction of anesthesia state dropped to zero near 29°C, at this temperature heat coma state of fish could be found.

This result indicated that NMDA receptor was temperature sensitive and cooperation between temperature and ketamine on anesthesia induction is effective. But the cooperation between temperature and volatile anesthetics (ethanol) on anesthesia induction is weak.

7. Protein Flexibility is Related to its Functioning

The experiment results show that natures of protein flexibility of a protein matches up well to its function. Natures (or distribution curve of active conformation of receptor or activity curve of receptors) of NMDA (N-Methyl-D-aspartate) receptor and GABA (gamma-aminobutyric acid) receptor is shown in Figure 7.

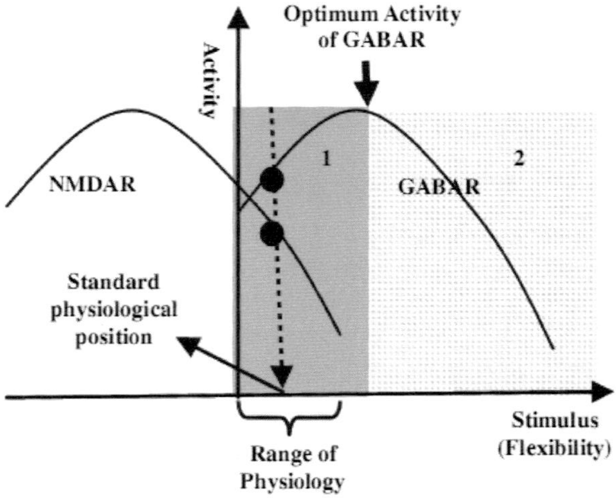

Figure 7. Natures of protein flexibility of the NMDA and GABA receptor.

The distribution curves of the active conformation of NMDA and GABA receptor are shown in this Figure. In the range of physiological conditions, a change of protein flexibility can result in opposite effects for the NMDA and GABA receptor respectively.

In this Figure, we have learnt that a change of protein flexibility has the opposite effect on the activity of different receptors. In the range of physiological conditions, a change of protein flexibility decreases the activity of NMDA receptor and increases the activity of GABA receptor.

In vivo, the excitability of a cell is mainly controlled by the NMDA (N-Methyl-D-aspartate) receptor and the GABA (gamma-aminobutyric acid) receptor. The NMDA receptor enhances the excitability while the GABA receptor lowers the excitability of a cell. At high temperatures or in the presence of protein denaturants, the excitability of a cell is increased and less activity of NMAD receptor is required, but more activity of the GABA receptor is needed; thus, natures of protein flexibility of NMDA or GABA receptors match well up with the requirement of a cell in a changing environment.

The positive and negative regulation of cell signaling by a change of protein dynamics (or flexibility) should be common mechanism for the regulation of signaling activities of a cell.

References

1. Zhao, Q. (2013). Nature of protein dynamics and thermodynamics. *Rev. Theor. Sci.* 1, 83–101
2. Brandts, J. F. and Hunt, L. (1967). Thermodynamics of protein denaturation. III. Denaturation of ribonuclease in water and in aqueous urea and aqueous ethanol mixtures. *Journal of the American Chemical Society*, 89(19), 4826–4838.
3. Kopito, R. R. (2000). Aggresomes, inclusion bodies and protein aggregation. *Trends in cell biology*, 10(12), 524–530.
4. Bennion, B. J. and Daggett, V. (2003). The molecular basis for the chemical denaturation of proteins by urea. *Proceedings of the National Academy of Sciences*, 100(9), 5142–5147.
5. Horwich, A. (2002). Protein aggregation in disease: a role for folding intermediates forming specific multimeric interactions. *Journal of Clinical Investigation*, 110(9), 1221–1232.
6. Rochet, J. C. and Lansbury, P. T. (2000). Amyloid fibrillogenesis: themes and variations. *Current Opinion in Structural Biology*, 10(1), 60–68.
7. Matveev, V. V. (2005). Protoreaction of protoplasm. *Cell Mol Biol*, 51(8), 715–23.
8. Johnson, C. R., Morin, P. E., Arrowsmith, C. H. and Freire, E. (1995). Thermodynamic analysis of the structural stability of the tetrameric oligomerization domain of p53 tumor suppressor. *Biochemistry*, 34(16), 5309–5316.
9. Seifert, R. and Wenzel-Seifert, K. (2002). Constitutive activity of G-protein-coupled receptors: cause of disease and common property of wild-type receptors. *Naunyn-Schmiedeberg's Archives of Pharmacology*, 366(5), 381–416.
10. Wang, A., Zhang, Z. and Zhao, Q. (2009). In vivo measurement of protein functional changes. *International Journal of Biological Sciences*, 5(5), 411.

Chapter 24

Signal Conduction and Neural Conduction

Summary: The unified model of signal conduction and the protein model of neural conduction are discussed in this chapter. We will show that neural conduction is not the propagation of electric signals, but the propagation of protein conformational signals. Thus, a systematic and unified knowledge of molecular biology, neuroscience, and physiology, has been successfully developed in this chapter.

Since we have studied the thermodynamic definition of biosignals in previous chapters, we will now discuss signal conduction mechanism in detail. The signal transduction activity of a cell occurs in a place, via the signal conduction pathway, where signaling activity can be transmitted into other parts of the body. The signal conduction mechanism plays a key role in the biological regulation of an animal.

In history, the concept of fluid and nerve regulation has been well established (1). When physiological regulation occurs at different places, a signal must be conducted into another place. Then the signal conduction is essential step for physiological regulation.

Previously experimental and diversified mechanisms of signal conduction, such as neural conduction, have been proposed on experimental observation, but there is no unified model for it. Although the signaling activity at molecular level is well known and well established in view of experimental science in field of biochemistry and molecular biology, the unified model for signal conduction has not been established theoretically.

In this chapter, we will discuss the unified model for signal conduction on the basis of the protein signal theory and we will show that the neural conduction is not the propagation of electric signals, but protein conformational signals.

1. Protein Conformational Couplings and Signal Conduction

According to signal theory of molecular biology we have discussed in previous chapter, any signal or signaling activity of molecular biology is related to one potherse or a thermal system within a protein (or protein conformational state), thus signal conduction is a propagation of protein conformational change or the spreading of signaling activity in space. The conduction or transduction of signals among different protein molecules can be carried out via: (1) a common ligand, if two proteins share the same ligand and protein conformational change linkage between these proteins can be carried out by this common ligand, such as cAMP, Ca^{++}, transmitter of nervous; (2) protein conformational mechanical coupling, which has been well studied in field of biochemistry; (3) common physical variables, such as electric field, mechanical force; if two proteins are both sensitive to an electric field, their conformational change can be coupled via the electric field.

A simple combination of these linkages of protein conformational coupling (potherse coupling) produces four basic types of biosignal conduction mechanisms with velocities greater than $1\,\mu m/s$. The ligand linkage of potherse changes can cooperate with the cell and vehicle and it results in mechanisms 1 and 2. The linkage of conformational coupling can occur alone and it will be discussed in mechanism 3. The linkage of potherse change by an electric field only occurs in mechanism 4 and it must cooperate with other mechanisms. We will discuss these mechanisms in detail as follows.

2. Mechanism #1: Sequential Ligand Release Signal Conduction

The best-known example of this type of signal conduction is Ca^{++} waves (2). The free Ca^{++} induces the release of Ca^{++} from the

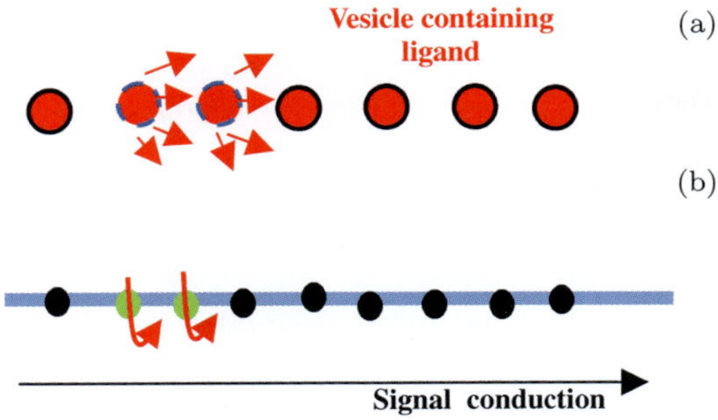

Figure 1. Two pathways of signal conduction of sequential ligand releasing mechanism.

In Figure 1a, the ligand is stored in small vesicles within a cell. After the ligand is released from vesicle, it diffuses into other vesicles and stimulates the release of ligands from other vesicles. The signal conduction can be then conducted into other area of the body. In Figure 1b, the cell membrane has channels which permit the release of ligands, when ligand is released out of cell membrane, it diffuses into other sites and stimulates the channel activity, step by step, which conduct the signal in other parts of the cell. The second mechanism (Figure 1b) has not been discovered by experimental biologists.

Ca^{++} stored in vesicles through Ca^{++} sensitive proteins. As this process repeats sequentially by diffusion, the Ca^{++} biosignal can transmit forward. If the time delay in transmitter diffusion between two adjacent cells (or vesicles) is 0.05 (100 Å) $\sim$ 5 ms (200 Å) (time delay of nerve transmitter), the velocity will be about 4–200 m/s. This type of signal transmission occurs in the cell or in the cell membrane. Another example of this type of signal conduction is the spreading of cell depression within a tissue (3). Two different pathways are illustrated in Figure 1.

3. Mechanism #2: Channel Conduction or Cell Communication

Biosignal conduction between cells plays an important role in the development and maintenance of cell functions (4–5). The time delay for transmitter diffusion between two adjacent cells is 0.1 $\sim$ 5 ms (this value is well known for the diffusion of neural transmitter between synapses). We further assume that cell length is 2–10 mm. When the time delay for conduction of the biosignal in the cell is neglected since it travels very fast via the protein conformational coupling mechanism, we can conclude that the signal transmission velocity can be estimated to be $0.5 \sim 100\,\text{mm/s}$. If the transmitter were a larger molecule (if it had such material), the conduction velocity would be slower than the mentioned. The signal conduction of this mechanism within our skin can be felt by us and ancient Chinese have studied it for a long time and termed it as channel conduction.

4. Mechanism #3: Protein Conformational Coupling

The protein conformational coupling mechanism for signal conduction occurs on the protein fibers (protein complex). The neural conduction along on demyelinated fibers is via the standard protein conformational coupling conduction. Neural conduction has been wrongly described as electrical signal conduction in history (see the following discussion).

5. Mechanism #4: Signal Conduction on Myelinated Nerve Fibers

Signal conduction on myelinated nerve fibers is due to a specific mechanism. The linkage of potherse change at the two sides of the sheath areas of myelinated fibers is carried out by an electrical field, so the effect of the electric field is essential in this case. In this case our results agree with the traditional theory of neural conduction in most aspects (6) (see the following discussion).

6. The Excitability of a Nerve Cell or Weiss Equation

The excitability of nerve cell membrane has been well studied for a long time. The Weiss equation, which describes the relationship among stimulus duration and strength, has been proposed too (7). The Weiss equation (expressed in Louis Lapicque's form) is

$$\mathrm{I} = \mathrm{b}(1 + \mathrm{c}/\mathrm{d})$$

Where I is current, b relates to the rheobase value and c relates to the chronaxie value over duration d.

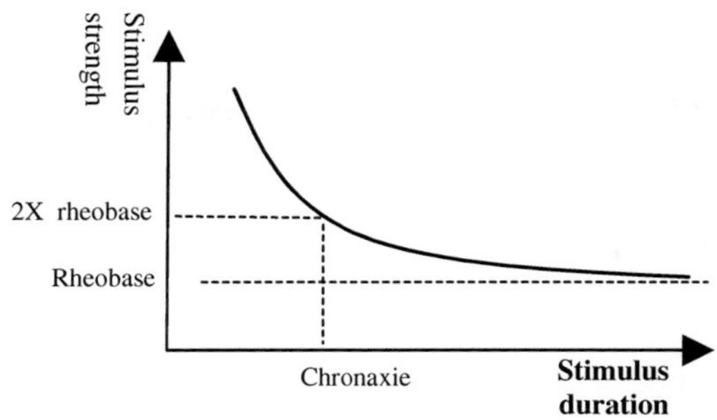

Figure 2. The excitability curve of a nerve cell.

In this Figure, Rheobase and chronaxie are points defined on the strength-duration curve for stimulus of an excitable tissue.

Or

$$V = V_a + W_b/t$$

Where V is voltage, V_a is rheobase (in voltage), W_b is constant.

The general properties of Weiss equation or excitability of nerve cell can be seen in Figure 2.

According to the protein theory, the open and closed state of a channel protein represents different potherse, the excitability of a nerve cell responding to a stimulus represents the mechanics characteristics of pothersal change of channels induced by electrical force. This can be analyzed by protein theory. The protein conformational change in an electrical field (or electrical force) can be described as:

$$dC = kV$$

Where k is the tensile strain constant, dC represents the protein thermodynamic conformational parameter change induced by the electrical field and V is the potential.

For the initial rate of protein conformational change toward protein structural change:

$$dC = g(V - V_0)dt$$

Where g is a constant and V_0 is the minimum potential required for static protein thermo-dynamic structural changes.

Then

$$\int_{C_0}^{C_0+\Delta C} dC = \int_0^T g(V - V_0)dt$$

$\Delta C = gVT - gV_0T$
$V = V_0 + \Delta C/(gT)$

Where C is the minimal protein conformational change between two different potherse, C_0 is the protein conformational parameter (for the closed state) and $(C + \Delta C)$ represents another protein conformational parameter of the potherse (open state).

By comparing it to the Weiss equation, we can get following conclusion

$$V_a = V_0 \quad W_b = \Delta C/g$$

The properties of protein conformational change of channel are shown in Figure 3.

In this Figure, we can see that rheobase represents minimal quantity of stimulus to induce potherse change (from closed to open state). The open state or closed state can also permit protein conformational change to some degree without inducing a change in state.

7. Biophysical Features of Protein Conformational Coupling

Protein conformational coupling occurs through a protein net or protein fiber. In order to deduce the biophysical features of standard protein conformation coupling and its velocity, we hypothesized that:

(1) The protein net (or protein fiber) is composed of the same protein complex. The diameter of this protein complex is about 70 Å. It is the average diameter of an enzyme complex.
(2) Although a protein is composed of many potherses (or protein conformations), two potherses (for a channel protein: open and closed states) are directly involved in signal conduction, and the

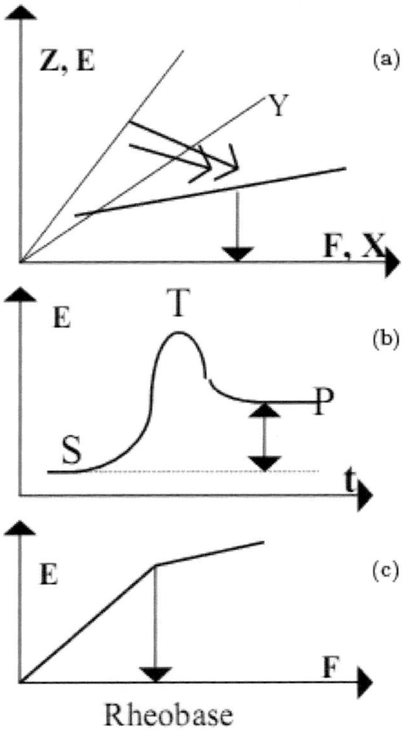

Figure 3. Features of Protein conformational change.

Figure 3a describes features of protein conformational change of a channel protein under voltage. Figure 3b describes pathway of protein conformational change induced by a change of temperature. Figure 3c represents simplest cases for Figure 3a. (Reproduced from Ref. 8. with permission)

conversion between these two potherses initiates signal transduction activity. Therefore, we can analyze signal transmission based on these two physical states.

(3) Energy required for the potherse change is supplied by signal transduction activity.
(4) The mechanical and thermodynamic properties of protein can be separated. A protein's mechanical stimulus can induce the deformation of a potherse but cannot make changes to a potherse directly. The potherse change results from inner thermal motion because the reorganization of the protein three-dimensional

structure is needed. Thus, protein conformational coupling is carried out by two physical mechanisms: mechanical coupling and thermodynamic coupling. Therefore, we may ignore mechanical coupling when we estimate the rate of protein conformational coupling. This hypothesis enables us to discuss questions within a simple model.

In fact, the electrotonic conduction known in neuroscience (6) is the conduction of conformational coupling.

(5) We have leant from the protein thermodynamic structure theory that the maximum rate of potherse change is about $4X10^8$.

Then, the maximum velocity of signal conduction for protein conformational coupling should be:

$$70\ (\text{Å}) \times 10^{-10}\text{X}4.2\text{X}10^{8} = 2.9\,\text{m/s}$$

The velocity of neural conduction of demylinated fibers is about 0.1–2 m/s. (6)

Considering that many different factors can influence signal conduction and that it is very difficult to discuss this theoretically, our estimation agrees well with experimental data. When the protein fiber has a very high tension, the velocity of signal conduction will become very fast. The signal conduction along the cardiac muscle cells is carried out by a protein conformational coupling mechanism (theoretically, other mechanisms can be easily excluded).

8. Nerve Cell Membranes Have a Solid Protein Net

If we hypothesize that neural conduction is carried out by the protein conformational coupling mechanism, we should demonstrate the existence of a protein fiber or protein net within a cell. In fact, many types of protein fibers have been well documented in cell biology, such as the cell microtubule, or microfilament. But here we will discuss other types of protein fiber or net exists in nerve cell membranes.

Much evidence supports the existence of a continuous solid protein net in or near nerve cell membranes. This protein net, consisting of receptors, channels, G proteins and other functional proteins, directly links to the neurotubule actincytosketon and synaptic secretary vesicles, (9–13), and sustains nerve cells morphologically and physiologically. Thus neural signals can be transmitted along it. The most powerful evidence comes from neuroscience. It is well known that electrotonic conduction can be transmitted within the nerve cell membrane (6). According to protein mechanistics, this represents the transmission of mechanical waves (or solitons) along the continuous protein net. The physical and physiological interaction between receptors and the cell skeleton can be clearly seen in spine dynamics, swelling, or mechanical excitability of nerve cells (14–19). The effect of the cell skeleton and neurofilament on neural conduction stability has been reported (20–21). The neural conduction from the soma to the hillock via the actin skeleton has been confirmed as well (22, 23). In Figure 4, we have drawn up a diagram of a continuous protein net/fiber according to the current knowledge of cell biology. However, the detailed structure of other possible patterns of a protein net still need to be studied in future.

Several models of neural conduction related to the cell skeleton have been proposed (24, 25). Penrose has discussed the possible function of the neurotubule in neuroscience for human thinking (26). He believes that our thoughts can be logically traced back to

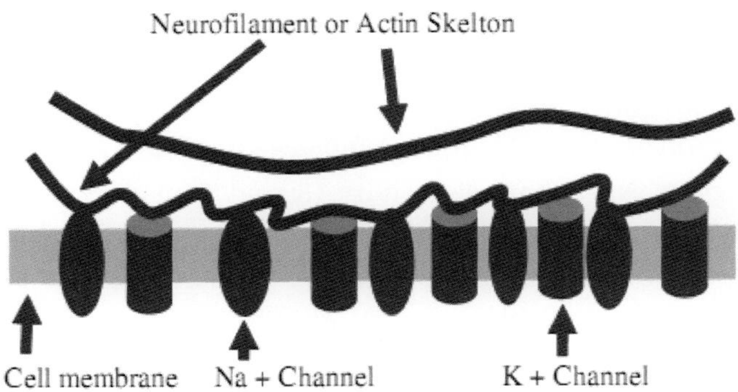

Figure 4. One pattern of protein net/fiber.

The pattern of protein net/fiber was drawn up according to our current knowledge of cell biology. The actin skeleton physically interacts with Na^+ and K^+ channel directly or via other proteins.

quantum states of the cell skeleton. Within our model, the quantum state of the cell skeleton is an unnecessary concept, but our thought is related to thermodynamics states of cell skeleton in some ways.

9. Neural Conduction of Demyelinated Nerve Fibers

As the velocity of signal conduction by protein conformational coupling mechanism is similar to that of neural conduction of demyelinated fibers, we must consider the possibility that it is in fact the working mechanism of neural conduction along demyelinated fibers. Hereafter we will show that this conclusion is the truth.

According to the electrical signal of neural conduction, the Na^+ channel is opened by depolarizing waves (electric field). However, in view of our theory, the Na^+ channels are opened by a mechanism of protein conformational coupling. Thence, the Na^+ channel opens just before action potential (AP). Thus, if we can confirm that the Na^+ channel opens just before action potential (AP), the old theory will be proved false.

Theoretically speaking, the AP experimentally measured is an admixture of a depolarizing wave and an electric field generated from ion flux. One way to distinguish these two types of potentials is by phase plotting dv/dt versus time (22, 23, 27, 28). see Figure 5(b).

We have drawn up the principles of neural conduction of demyelinated fiber in Figure 5.

The mechanical wave travels faster and deforms the protein net, but potherse change does not occur in this step. The potherse change (protein conformational coupling) follows after the mechanical waves and it triggers Na^+ channel activity (signal transduction activities) that then results in the generation of action potential. Many types of signal transduction activities may be triggered and most of them contribute nothing to action potential. Thus, the action potential is only the auxiliary product of signal transduction activities. It does not influence the signal conduction velocity of demyelinated nerve fiber.

As we will discuss bollow, the powerful evidence of our theory comes from the results obtained from studying of neural conduction about myelinated fibers.

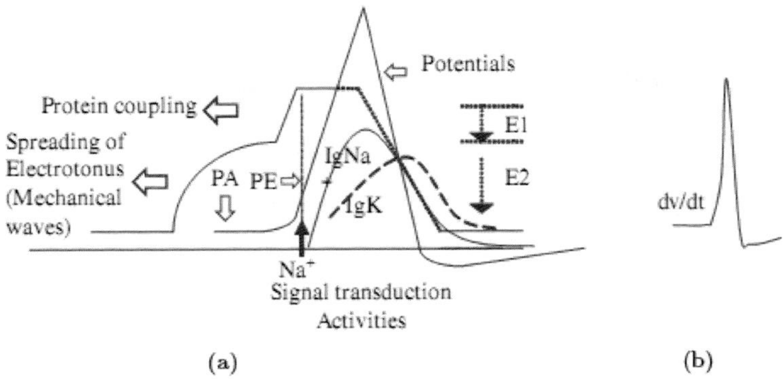

Figure 5. Principle of nerve signal transmission on nerve demyelinated fiber. (a) The energy level of protein net (voltage-gated Na+ channel) is drawn. The potherse change is ended completely when protein conformational coupling occurs. This process is then repeatedly sequential along the nerve fiber, allowing the transmission of neural signals. E1: energy transmitted by protein conformational coupling. E2: The energy transmitted by mechanical waves. PM: gating potential (current) generated from the protein conformational change transmitted by mechanical waves, considered by some to be electrotonic conduction, which is small. PI: The potential produced in the diffusion of ions. (b) Phase plot of dv/dt versus time.

10. Signal Transmission on Myelinated Fibers

Most human nerve fibers are myelinated fibers. As the protein conformational coupling cannot pass though sheath area of nerve fiber, the linkage of potherse change between different nodes of Ranvier must be carried out by an electric field. Unlike the protein conformational coupling mechanism, the electrical linkage of potherse changes is slow and time delay (or latency) can be observed. At present, it is uncertain if the mechanical coupling mechanism (along the cell skeleton) contributes to neural conduction. If it does, it is not the controlling step in neural conduction. Figure 6(a) illustrates its principle.

The conclusions of our theory agrees well with experimental results (6, 29–31).

Since temperature greatly affects protein mechanical properties (or protein flexibility), the time delay (TD) increases rapidly with decreasing in temperature; thus, the neural conduction velocity

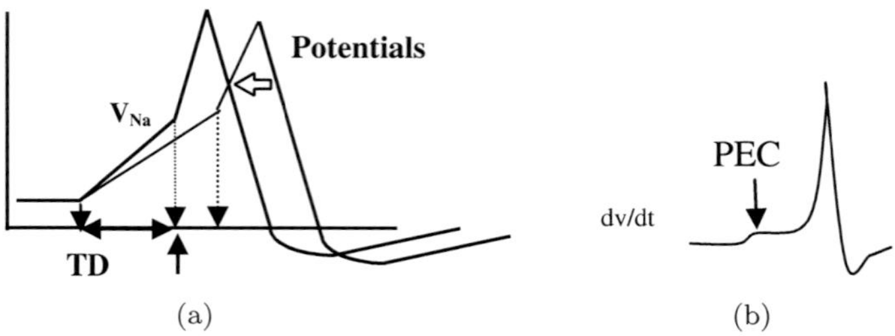

Figure 6. Principle of nerve signal transmission on myelinated fibres.

PEC: Potential generated from the electric circuit of action potential (or depolarizing wave) that produces the electrical linkage of protein conformational change between different nodes of Ranvier. TD: Time delay in the electrical linkage of potherse. VNa: action potential generated from Na^+ influx. b) Phase plot of dv/dt versus time.

become very much faster at high temperature (30). It is powerful evidence for our theory.

The velocity of neural conduction can be described by:

$$V = (L_1 + L_2) \div (DT + L_1 \div V_d)$$

Where L_1 is the sheath length, L_2 is the length of node of the Ranvier, DT is the time delay in the electrical linkage of protein thermodynamics-system, V_d is the neural conduction velocity along the demyelinated fiber and V is the velocity of myelinated fiber.

In Figure 6(a), the DT represents the time delay required for potherse change induced by an electric field. There is no DT segment in Figure 5(a). This denotes a mechanism difference between neural conduction mechanisms of demyelinated and myelinated nerve fibers.

11. Synapse Transmission

According to the protein model of neural conduction, synapse transmission is carried out by a transmitter. The receptor conformational change induced by neural transmitter that can suppress (for inhibitor) or stimulate (for activator) the activation of the Na^+ channel in many different ways. The IPSP (inhibitory postsynaptic potential) and EPSP (excitatory postsynaptic potential) are only one product of receptor activity (ion channels). Many signal transduction pathways, which do not induce ion flux, cannot be measured electrophysiologically. In some cases, GABA (inhibitory neurotransmitter) cannot induce ion flux, and the IPSP cannot be observed, but inhibition still occurs (32, 33).

The volatile general anesthetics and many medicines can also influence synapse transmission, but the EPSP and IPSP cannot be observed (34, 35).

12. Biosignal Integration Mechanism at the Cytoskeleton

The channel protein binds to the cell skeleton and this interaction regulates the excitability of nerve cells and neural conduction. Although our understanding about it is poor, some powerful evidence was observed (22, 23). The AP is generated in the axonal initial segment (AIS), although the stimulus is applied at the synapses or soma (36). What is the working mechanism underlying this phenomenon? The study of Stuart sheds light on this (22). He showed that the cell skeleton played a role in the transmission of a signal from soma to AIS. Evidences are:

(1) the AP is normally initiated in AIS, rather than in soma, which can be prejudged by the fact that the AP of the AIS precedes the AP of the soma;
(2) If the AP of the AIS is suppressed by TTX, a blocker of the Na^+ channel, the AP of the soma precedes the AP of the AIS;
(3) if the actin skeleton is partly dissociated, the AP of the soma again precedes the AP of the AIS.

These tests firmly indicate that the cell skeleton is involved in the initiation of neural conduction. The detailed mechanism is shown in Figure 7.

Although this mechanism cannot account for all the properties of neural initiation, it can certainly describe some features of it. In addition, it provides an alternative mechanism for synaptic integration (37). The blocker of the K^+ channel can regulate the

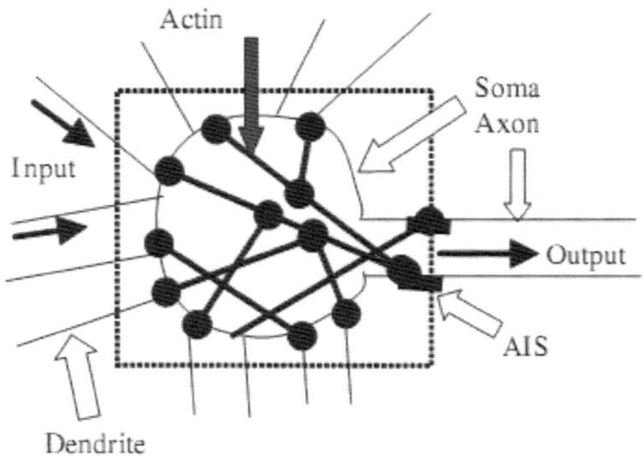

Figure 7. Bioinformation is integrated at the soma, the actin-skeleton, and the AIS.

The cell skeleton binds channel proteins and mechanical signal from the soma to the AIS can be transmitted along it. So the soma, cell skeleton and AIS act as neural signal processor in vivo. The soma and dendrite receive numerous synapse signals, which can be transformed into the dynamic nature of cell skeleton and finally into the normal AP in AIS. The AP can be then back-propagating into soma and dendrite.

excitability of nerve cells and can change the firing fashion of a nerve cell (35). However, adding a blocker of the K^+ channel at an apical area of dendrites can change the excitability of the soma and alter the firing fashion (38). How can we explain this with an electrical signal theory of neural conduction? We can only accept the idea there is a mechanical relationship between the dendrites and the soma.

Theoretically speaking, the change of fashion of informational integration induced by the reorganization of cell skeleton involved in neural signal transmission at the soma area can account for our thinking and memory.

13. Comparisons between Protein Model and Electric Model for Neural Conduction

There are many different understandings between the protein model and the electrical signal of neural conduction. In Table 1, we have classified different terms used by different theories.

Our theory confirms the concept of the separation between signal initiation and signal conduction. The duration of the AP

Table 1. Different scientific terms for two models of neural conduction.

Electric model for neural conduction	Protein model for neural conduction
Electric signal	Protein conformation signal
propagation of electric signal	Propagation of protein conformational signal
Positive propagation	Mechanical coupling among proteins
Active propagation of electric signal	No such concept
Electric excitability of membrane	Excitability of a cell to stimulus
Protein polarization in electric field	Protein conformational change induced by mechanical force applied on protein by electric filed
Na^+ channel is activated by electric field in neural conduction	Na^+ channel is activated by protein conformational coupling in neural conduction
Rheobase	Minimum strength of electric field required for the firing of a nerve cell
Hyperpolarization	Unfavorable protein conformation of channels to be activated
Depolarization	Favorable protein conformation of channels to be activated

is independent of signal initiation and can be very long (32). The signal (neural) initiation is carried out by the activation of the Na^+ channel and many ways of signal transduction can be induced by an AP, which results in many types of changes, both physical and biochemical, within the cell membrane (39, 15–18). The complex relationship among these different types of biosignals can be altered, which will induce reorganization of the protein net as well as nerve cell excitability, and finally result in the production of LTP and LTD (40).

According to our theory for neural conduction, the nerve cell may have different types of protein nets (or protein fibers) and different types of neural signals which can be transmitted via different protein nets (or protein fibers). It has been observed that two different neural conduction mechanisms (molecular), Ca^{++} and Na^+ action potentials, can be carried out by the same nerve cell (41).

References

1. Kandel, E. R., Schwartz, J. H. and Jessell, T. M. (Eds.). (2000). Principles of neural science (Vol. 4, pp. 1227–1246). New York: McGraw-Hill.
2. Jaffe, L. F. (1993). Classes and mechanism of calcium waves. *Cell Calcium.* 14, 736.
3. Somjen, G. G. (2001). Mechanisms of spreading depression and hypoxic spreading depression-like depolarization. *Physiol. Rev.* 81, 1065.
4. Zhao, Q. (1996). The physical movement of macromolecules and the physical process of signal transduction. *Chinese J of Rehabilitation Theory & Practice*, 2, 135.
5. Barbe, M. T., Monyer, H. and Bruzzone, R. (2006). Cell-cell communication beyond connexins: the pannexin channels. *Physiology*, 21(2), 103–114.
6. Mountcastle, V. B. (1980). Medical physiology (Vol. 1 14th Ed) C.V. Mosby Company, London.
7. Irnich, W. (2002). Georges Weiss' fundamental law of electrostimulation is 100 years old. *Pacing and clinical electrophysiology*, 25(2), 245–248.
8. Zhao, Q. (2001). Irreversible thermodynamics theory for protein folding and protein thermodynamics structure. *Prog Biochem Biophys.* 28, 429–435.
9. Chen, X., Winters, C., Azzam, R., Li, X., Galbraith, J. A., Leapman, R. D. and Reese, T. S. (2008). Organization of the core structure of the postsynaptic density. *Proc. Natl. Acad. Sci. USA.* 105, 4453.
10. Mazzochi, C., Benos, D. J., Smith, P. R. (2006). Interaction of epithelial ion channels with the actin-based cytoskeleton. *Am. J. Physiol. Renal. Physiol.* 291, 1113.
11. Cingolani, L. A. and Goda, Y. (2008). Actin in action: The interplay between the actin cytoskeleton and synaptic efficacy. *Nature Rev. Neurosci.* 9, 344.
12. Luna, E. J. and Hitt, A. L. (1992). Cytoskeleton-plasma membrane interactions. *Science.* 258, 955.
13. Ananthakrishnan, R. and Ehrlicher, A. (2007). The forces behind cell movement. *Int. J. Biol. Sci.* 3, 303.
14. Fischer, M., Kaech, S., Knutti, D. and Matus, A. (1998). Rapid actin-based plasticity in dendritic spines. *Neuron.* 20, 847.
15. Iwasa, K., Tasaki, I. and Gibbons, R. C. (1980). Swelling of nerve fibers associated with action potentials. *Science.* 210, 338.
16. Tasaki, I. and Byrne, M. (1990). Volume expansion of nonmyelinated nerve fibers during impulse conduction. *Biophys. J.* 57, 633.

17. Iwasa, K. and Tasaki, I., (1980). Mechanical changes in squid giant axons associated with production of action potentials. *Biochem. Biophys. Res. Commun.* 95, 1328.
18. Tasaki, I. (1999). Rapid structural changes in nerve fibers and cells associated with their excitation processes. *Jpn. J. Physiol.* 49, 125.
19. Tasaki, I. and Iwasa, K. (1982). Further studies of rapid mechanical changes in squid giant axon associated with action potential production. *Jpn. J. Physiol.* 32, 505.
20. Kriz, J., Zhu, Q., Julien, J. P. and Padjen, A. L. (2000). Electrophysiological properties of axons in mice lacking neurofilament sub.unit genes: disparity between conduction velocity and axon diameter in absence of NF-H. *Brain Res.* 885, 32.
21. Klose, M. K., Armstrong, G. and Robertson, R. M. (2004). A role for the cytoskeleton in heat shock-mediated thermoprotection of locust neuromuscular junctions. *J. Neurobiol.* 60, 453.
22. Kole, M. H. P., Ilschner, S. U., Kampa, B. M., Williams, S. R., Ruben, P. C. and Stuart, G. J. (2008). Action potential generation requires a high sodium channel density in the axon initial segment. *Nature Neurosci.* 11, 178.
23. Kole, M. H. P. and Stuart, G. S. (2008). Is action potential threshold lowest in the axon. *Nature Neurosci.* 11, 1253.
24. Rvachev, M. M. (2003). Alternative model of propagation of spikes along neurons. http://arXiv:physics/0301063v2 [physics.bio-ph] [internet]
25. Heimburg, T. and Jackson, A. D. (2005). On soliton propagation in biomembranes and nerves. *Proc. Natl. Acad. Sci. USA* 102, 9790.
26. Penrose, R. (1989). The Emperor's New Mind. Oxford Press, Oxford
27. Coombs, J.S., Curtis, D. R. and Eccles, J. C. (1957). The generation of impulses in motoneurones. *J. Physiol.* 139, 232.
28. Yu, Y., Shu, Y., McCormick, D. A. (2008). Cortical action potential backpropagation explains spike threshold variability and rapid-onset kinetics. *J. Neurosci.* 28, 7260.
29. Frankenhaesuer, B. and Huxley, A. F. (1964). The action potential in the myelinated nerve fibre of Xenopus Laevis. As computed on the basis of voltage clamp data. *J. Physiol.* 171, 302.
30. Frankenhaesuer, B. (1965). Computed action potential in nerve from Xenopus Laevis. *J. Physiol.* 180, 780.
31. Hutchinson, N. A., Kolesand, Z. J. and Smith, R. S. (1970). Conduction velocity in myelinated nerve fibres of Xenopus laevis. *J. Physiol.* 208, 279.
32. Hammond, C. (2008). Cellular and molecular neurophysiology. 3rd ed. Academy press, London.

33. Bear, M. F., Connors, B. W. and Paradiso, M. A. (1996). Neuroscience, exploring the brains. Williams & Wilkins, London.
34. Butterworth, J., Raymond, S. and Roscoe, R. F. (1989). Effects of Halothane and enflurane on firing threshold of frog myelinated axons. *J. Physiol.* 411, 493.
35. Yue, C. and Yaari, Y. (2006). Axo-somatic and apical dendritic Kv7/M channels differentially regulate the intrinsic excitability of adult rat CA1 pyramidal cells. *J. Neurophysiol.* 95, 3480.
36. Colbert, C. M. and Johnston, D. (1996). Axonal action-potential initiation and Na channel densities in the soma and axon initial segment of subicular pyramidal neurons. *J. Neurosci* 16, 6676.
37. Spruston, N., (2008). Pyramidal neurons: dendrite structure and synaptic integration. *Nature Rev. Neurosci.* 9, 206.
38. Beck, H. and Yaari, Y. (2008). Plasticity of intrinsic neuronal properties in CNS disorders. *Nature Rev. Neurosci.* 9, 357.
39. Chiang, E. and Strowbridge, B. W. (2007). Diversity of neural signals mediated by multiple, burst-firing mechanisms. *J. Neurophysiol.* 98, 2716.
40. Malenka, R. C. and Bear, M. F. (2004). LTP and LTD: An embarrassment of Riches. *Neuron.* 44, 5.
41. Cavelier, P., Pouille, F., Desplantez, T., Beekenkamp, H. and Bossu, J. (2002). Control of the propagation of dendritic low-threshold Ca2+ spikes in Purkinje cells from rat cerebellar slice cultures. *J. Physiol.* 540, 57.

Chapter 25

Signaling Network Theory and System Biology

Summary: This chapter discusses basic and thermodynamic principle of the biosignal network theory or system biology. In view of the signaling theory, a physiological system that represents a specific signaling network which is thermodynamically stable and shows independence to some degree. The properties of signaling network could be quantitatively described by the law of partition function of a thermal system. Some biological phenomena, such as plant vernalization, canceration, sexualization, and oscillation, are also highlighted in this chapter. (This chapter is a copy of the paper: A Thermodynamic Model of Protein Signaling Network. Reviews in Theoretical Science 4, pp. 1–9, 2016, with minor modification).

Cells and living bodies comprise of complex physiological systems. Although the general properties of biological organisms can be simply described based on the basic principles of system biology, scientists have difficulty in explaining the basic principles of the physiological system based on the principle of thermodynamics (1, 2).

Several scientific questions must be addressed to obtain a comprehensive understanding of system biology. First, the signaling network and its capability to represent a thermodynamic system must be defined clearly. Second, the system concept and its logical system must be given a scientific definition, one that is applicable in the field of system biology. Third, the biosignal concept must be given a thermodynamic definition that can serve as a theoretical foundation for discussing the signaling network model. Fourth, a thermodynamic principle that covers the thermal properties of a system must be identified.

Other scientists use experimental scientific methods to address the aforementioned questions. A number of scientists have outlined thorough signaling pathways of the biological system or signaling network. Some scientists have constructed a signaling network by adopting the dynamic system model which properties can be simulated (4-15). Although the pursuits of these scientists have produced valuable findings, they have ignored the general principles of the biological system, which hinders a comprehensive understanding of system biology. Although the signaling network and physiological system are two well-known concepts, the relationship between these concepts cannot be theoretically interpreted (2, 3, 11).

The system concept, pure system logic, thermodynamic definition of biosignal and partition function of a thermal system have been recently developed to investigate the irreversible thermodynamics and protein thermodynamic structure theory (3, 16–30). Three concepts has also been developed (i.e., any biological function corresponds to one biosignal network, any signaling network represents one thermodynamic system, and the thermodynamic state of core proteins in the signaling network can control other properties within the network) as bases for the development of the thermodynamic model of the signal network, which will reveal the general principles of the signal network or system biology and also establish a logical relationship between biological function and protein thermal motion.

This paper also addresses other related biological phenomena, such as canceration, sexualization, wheat vernalization, and biological oscillation.

1. Basic Facts and Core Structure of Protein the Signaling Network

Physiological changes result from molecular reactions. In the protein biosignal model, any molecular change or reaction is related to a protein, such as an enzyme. The thermodynamic changes in the physiology of a living body or a cell are induced by the conformational or pothersal changes in the protein (16–20). Therefore, the molecular changes can be represented by the thermodynamic properties of the protein. According to the protein signal theory, a molecular change is induced by the signaling activity of a protein. Therefore, these two concepts must be investigated by adopting the protein theory (16).

The fastening of a metabolite or nutrient, such as glucose, lipid, ATP, ADP, Na^+, K^+, Ca^{++}, and Mg^{++}, to proteins can induce a thermodynamic change in the signaling proteins. Physical (or quantum) factors, such as pressure, mechanical force, electron transition of the ligand of a signaling protein, can also induce a change of the thermodynamic state of signaling protein (21–23). Their effects of signaling activity are represented by and analyzed through the properties of the signaling protein (16). For example, a volatile general anesthesia can induce a thermodynamic change in the proteins of nerve cells, thereby reflecting the potency of the anesthesia (16, 20, 24).

Discrete physiological states are observed when a living body or a cell continuously responds to variable stimulus (7). According to the signal theory, these physiological states represent different signaling networks of a cell.

The changes in the physiological fashion of a signaling network are induced by the pothersal or conformational changes in a signaling protein, as represented by the activation of receptors. Figure 1 shows that a change in the signaling network can be processed either reversibly or irreversibly.

A reversible process is observed when the environmental factor acts upon a signaling protein (PS) and induces conformational change. An irreversible process is observed when a none-specific biological signal (NBS) acts upon a PS and induces conformational change. The product (PP) can be permanently perpetuated if PP acts upon a specific biological signal with a signaling activity similar to that of NBS. Such an effect can be observed in the activation of trypsinogen and prion biology (25, 26).

Numerous positive and negative feedback are regulated in vivo, which constructs the complex signaling network of a cell. In certain cases, the core structure of the signaling network comprises several

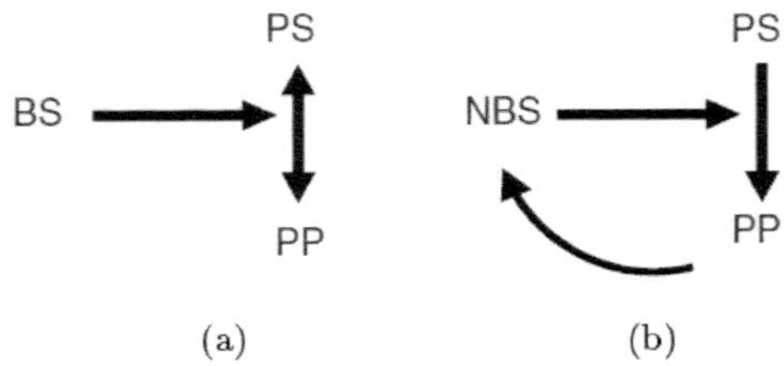

Figure 1. Basic fashions of protein conformational change.

BS is a biological signal, PS is a signaling protein, PP is a product of PS, and NBS is a none-specific signal. Figure 1a: Reversible processes. The conformational change of protein under a stimulus (or BS) is reversible. Figure 1b: Irreversible processes. The conformational change of PS can be induced either by NBS or PP. NBS initiates the signaling activity, which in turn stabilizes the network. The conformational change of protein can be either reversible or irreversible from the perspective of biochemistry.

proteins and the state of these proteins defines the properties and fashions of the signaling network. Most protein signals influence the quantitative properties of the signaling network but cannot determine the fashion of the network.

Based on the protein thermodynamic structure, a biosignal or molecular function corresponds to one potherse or a thermodynamic system of a protein (16). Unlike the molecular biosignal that corresponds to one thermodynamic system of a protein, the signaling network corresponds to one thermodynamic system of a core structure. Although such structures may vary from case to case, the core structure of a signaling network can be theoretically considered as one protein complex. Therefore, the general nature of the signaling network can be analyzed by applying the system theory and the protein thermodynamic structure theory without necessarily examining the nature of a signaling network.

These findings produce different interpretations of the signaling network concept within the scientific community. Experimental scientists define the cellular signaling network as a product of the mutual connections among biological molecules or molecular reactions and of the complex characteristics of such a network. However, this paper focuses only on a few protein species within a cell. They exhibit a certain degree of independence in their activities and can be

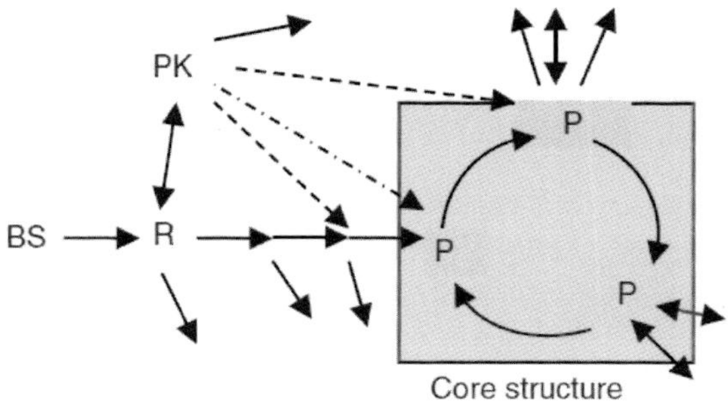

Figure 2. Core structure of a signaling network.

BS represents the environmental factors, PK is the protein kinase, and P is a protein. The core structure of any signaling network can be theoretically isolated, and certain properties of a signaling network can be represented and analyzed by the nature of their core structures. The proteins in the core structure can be connected to numerous proteins in vivo and to the response stimuli of the environmental factors.

theoretically isolated from many cellular activities. Some properties of a cell can be identified by analyzing the thermodynamic properties of the core protein. Figure 2 outlines the general structure of a cellular signaling network.

The core structure of a signaling network related to plant vernalization and cancer is the protein–chromosome complex or chromatin, which controls the gene expression or transcriptional activity. DNA methylation influences the structure of DNA or chromatin (27–30) as well as the transformation of the signaling network. The cells involved in anesthesia and prion disease have receptor cell skeleton complex (16, 24) and the prion protein (26) as their core structures, respectively. A biological trait or function corresponds to one signaling network. Cancer, disease state, anesthesia, diabetes, cell types of animals, and phases in the cell cycle are among the examples of biosignal networks (31, 16, 24, 32, 33). Apoptosis and vernalization are among the transformation processes that occur between different signal networks (34–36). Anesthesia represents the simplest case of a signaling network. According to the thermodynamic model, the transformation from an anesthetic state to a waking state can be reversed, whereas the transformation from a normal cell to a cancerous cell cannot be reversed. The signaling networks show a hierarchical structure in vivo.

2. Thermodynamic Stability and Distribution Curve of the Biosignal Network

Thermodynamic stability is a basic concept in a thermal system. As numerous thermal systems are transformable, the distribution curves of thermal systems can be measured experimentally and deduced from the partition function concept (19).

Even though only a few studies have examined the thermodynamic properties of a signaling network, this paper has obtained valuable insight with regard to such topic. The proposed thermodynamic model accounts for many different properties of a signal network. This paper refers to wheat vernalization as an example, which is the earliest-discovered example of a signaling network (35–36).

Figure 3 shows the typical stability curve of signaling networks based on the partition function of a thermal system.

Because a living body comprises numerous, complex signaling networks, the tolerance of the living body represents the stability of the signaling networks. Such representation has been described in the field of biology by Shelford's law of tolerance (37).

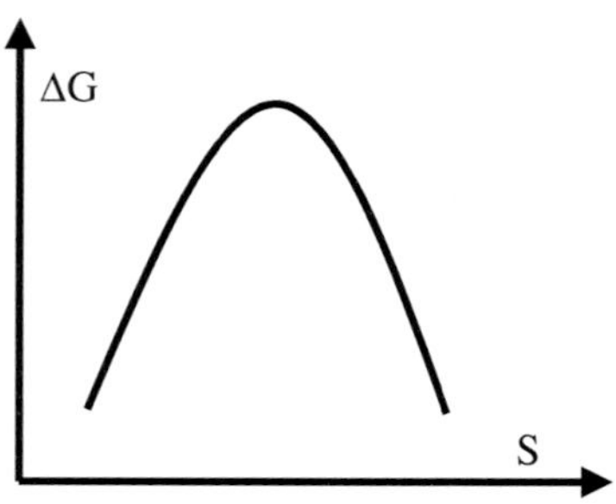

Figure 3. Typical stability curve of a network.

S represents the strength of the stimulus. This type of stability of a signaling network appears when a network is transformed among three or more types of signaling networks.

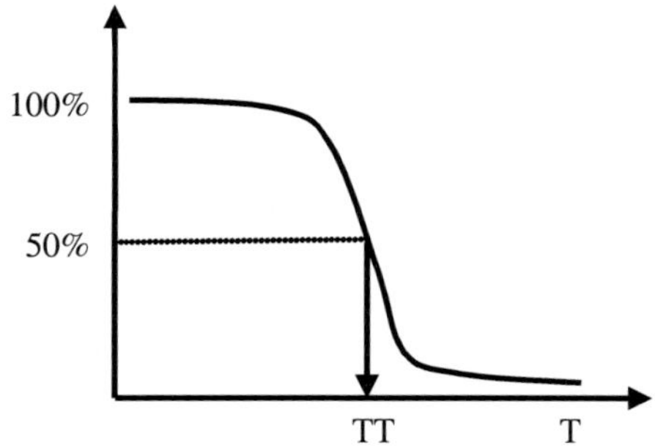

Figure 4. Distribution curve of signaling networks between two networks. T represents the environment temperature, and TT represents the transition temperature between vernalized and unvernalized wheat.

Figure 4 illustrates the distribution curves of the two signaling networks, namely, the vernalization of plants and production of anesthesia.

Given the limited data on thermodynamics, the distribution curve and other thermodynamic parameters cannot be obtained experimentally. The transition stimulus (i.e., the transition temperature for plant vernalization) is an acceptable measure for the stability of the signaling network. For example, even though the distribution curve of vernalization cannot be measured, the transition temperature (TT) can be precisely measured to obtain a general profile of stability (38). The stability of the signaling network is reflected in the distance between the physiological condition and transition stimulus (i.e., the farther the physiological condition from the transition stimulus, the more stable the signaling network becomes). In medical science, the stability of a normal physiological state to a disease is called the susceptibility of a disease.

3. Dynamic Nature of Signaling Network Transformation

The transformation between two different signaling networks is a dynamic process, the nature of which can be described and analyzed based on the principles of a protein signaling network.

The dynamic nature of signaling network transformation is not empirically supported by any thermodynamic data. Therefore, experimental scientists have resorted to other methods, such as a timescale for plant vernalization, to analyze the dynamic nature of signaling network transformation. Such a timescale represents the dynamic nature of transformation between unvernalized and vernalized cellular signaling networks in plants (39, 40). Figure 5 shows the dynamic nature of wheat vernalization.

Two types of changes occur when a signaling network acts upon a stimulus. On the one hand, a quantitative change occurs when the measurement of the signaling network continuously changes whereas the fashion of the core structure remains unchanged. On the other hand, a qualitative change is observed when the core structure of the signaling network changes and when a new fashion of signaling network is produced. Figure 6 presents examples of these two changes.

These changes reflect the robustness of the signaling network (1). The activity of a cellular signaling network may induce a quantitative change in the connected signaling network rather than inducing its transformation. Therefore, numerous types of signaling network can be combined freely.

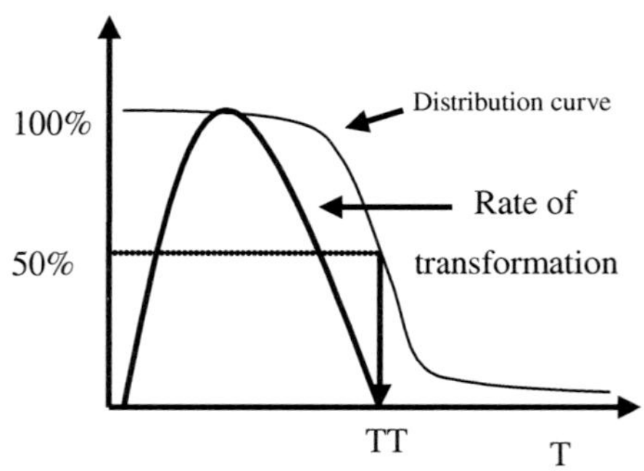

Figure 5. Dynamic stability of signaling activity.

The rate of wheat vernalization and the distribution curve of the vernalized wheat are presented in this figure. A slow metabolism rate is observed at low temperatures, which also slows down the wheat vernalization. As the metabolism rate increases, the dynamic stability decreases at any temperature.

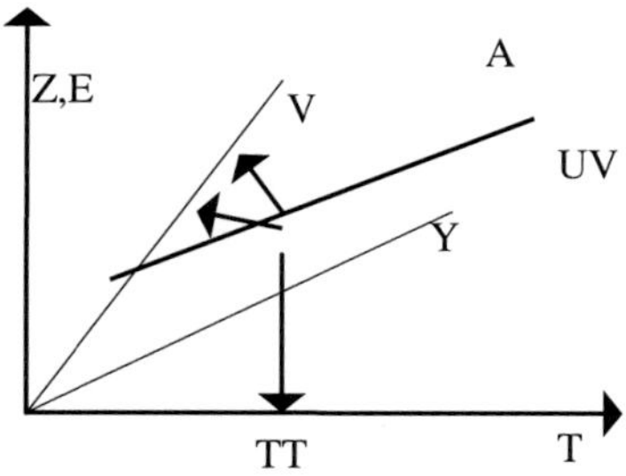

Figure 6. Dynamic nature of wheat vernalization.

E is the stability energy of the signaling network, T is the temperature, Y and Z are the two hypothesized parameters of signaling network, UV is the unvernalized state of wheat, V is the vernalized state of wheat, and TT is the transition temperature.

Most cellular signaling activities are accompanied by the activation of protein kinase cascade. Protein phosphorylation is fundamental in regulating the thermodynamic state and signaling activity of a protein. By modulating the sensitivity and strength of the cellular signaling network, the protein phosphorylation signal is necessary for the cellular signaling activity. However, this signal is neither the basic component of the cellular signal nor the main pathway from the extracellular signal to the cell nuclei (41).

The nature of the continuous change in the signaling network remains vague. Such change may represent the thermodynamic or dynamic natures of core proteins or may induce other types of changes in a cell. The proposed thermodynamic model cannot determine the actual nature of such changes. Both quantitative and qualitative changes may contribute to the transformation of the signaling network, the circumstances of which may vary from case to case.

4. Role of Gene Mutation in Signaling Network

Genes are the fundamental sources of biological information. Gene mutation can alter the protein sequence as well as the dynamic and thermodynamic natures of signaling protein. The proposed thermodynamic model indicates that some types of gene information are transformed into information on the thermal motion of a protein before they are integrated into the signaling protein.

The model proposes two mechanisms on the role of gene mutation in the signaling network. First, a gene mutation can change the shape of the distribution curve of a signaling network (D1 to D2), and then change the state of a signaling network. Second, a gene mutation shifts the distribution curve along the stimulus (D1 to D3), and then changes the state of a signaling network. Figure 7 demonstrates these two mechanisms.

These mechanisms become more pronounced when the distribution curves of the signaling networks of both mutations are measured. If the distribution curve cannot be measured easily, the transition point (TT for vernalization) can be used as a metric for the second mechanism. Theoretically, a gene mutation in a protein within the core structure of a signaling network can change the shape of the distribution curve, whereas a gene mutation in an associated protein changes the signal strength in the signal transduction and shifts the curve along the stimulus.

Wheat vernalization may vary among different wheat species that are generated through gene mutation (42). The mutation modifies the position of the distribution curve of vernalization, which improves the adaptation of the plant to different environments. As the signaling

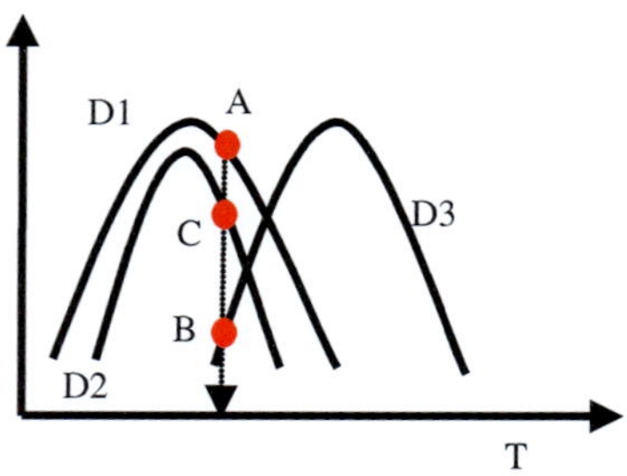

Figure 7. Changes in the state of a signaling network.

D1, D2, and D3 represent the two distribution curves; and A, B, and C represent the physiological states at specific conditions. The gene mutation can shift the physiological state from A to C by shifting the distribution curve from D1 to D2. A gene mutation can change the physiological state of the signaling network from A to B by shifting the position of the distribution curve.

network of the vernalized wheat has a stable core structure or chromatin, the model predicts that the gene mutation of these wheat species is induced by the second mechanism. In other words, most gene mutations in wheat vernalization occur in soluble proteins rather than in the proteins in chromatin.

The information on the effects of different genes or mutations is thermodynamically integrated within a signaling protein. Two types of information integration, namely, neutralization and cooperation, are widely observed in experimental science.

5. Thermodynamic Mechanism of Canceration

The canceration process is significant in medical and biological sciences. Its mechanism is discussed in this paper by adopting the aforementioned principles of signaling network. The thermodynamic model defines canceration as the transformation process of the cellular signaling network. Chromatin composes the core structure of the cancer signaling network and controls the mode of gene expression. Figure 8 illustrates the thermodynamic principle of canceration.

The canceration mechanism can be analyzed from two aspects, namely, the core structure of cell canceration and the physical formation of the core structure of the cancer signaling network. Given that cancer cells express various proteins that are not expressed in normal cells, the influencing factors on the structure and conformational state of chromatin become potent carcinogenetic factors under a specific condition. Certain carcinogenetic factors, such as X-ray, act upon the DNAs and proteins connected to chromatin.

When a cell acts upon a stimulus, the proteins of such a cell undergo a conformational change, which induces the conformation of numerous proteins. Some of these conformed proteins, particularly C-prone-PC (Figure 8), are prone to canceration. A cell that carries such a conformed protein may produce a cancer-like chromatin or cell. The cell may also remain in its normal state and may be distributed to two types of cells, as described by the law of partition function (19). Therefore, the cancer-like cell may further transform

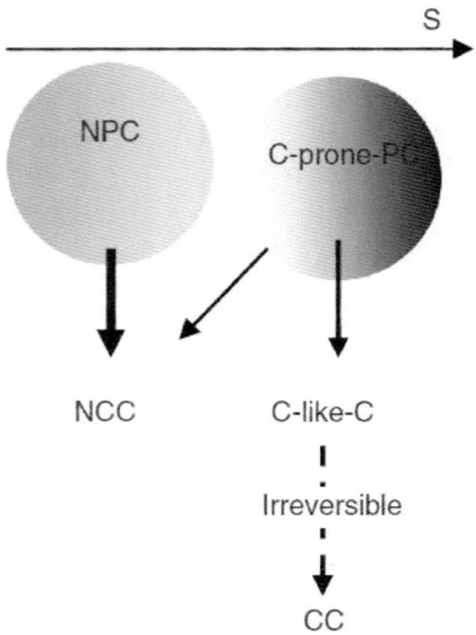

Figure 8. Thermodynamics model for cancer induction.

S is the stimulus, NPC is the normal protein conformation, C-prone-PC is the cancer-prone protein conformation, NCC is the normal cell chromatin, C-like-C is the cancer-like fashion of chromatin, and CC is the cancer fashion chromatin. When a normal cell acts upon a stimulus, the conformational space (NPC) shifts to C-prone-PC, thereby forming a conformational space similar to that of a cancer cell, an NCC, and an abnormal cell chromatin (C-like-C).

into a cancerous cell, which cannot be reversed. The gene mutation, metal ion (i.e., Ni^{++} and Mn^{++}), protein phosphorylation, and DNA play significant roles in the conformational changes of protein, which induce the formation of chromatin. These factors can also shift the distribution curve of normal and cancerous cells. Most gene mutations in cancerous cells occur in kinase genes and DNA-binding proteins used in the remolding of chromatin (31). Different types of canceration signals or factors, such as carcinogenic factors, anti-cancer drugs, extracellular matrices, and microenvironmental factors, are integrated in the last step of the chromatin formation (43–45).

The following features of canceration are deduced from this mechanism: 1) numerous types of gene mutations have been identified through experiments, 2) no single gene mutant can account for all canceration types or the effect of one gene mutant on cell canceration can be produced by another gene mutant, 3) the effects of gene mutants on cell canceration can be integrated, and 4) based on molecular biology, cancer can be induced by completely different gene mutations or, in other words, the canceration is a heterogeneous mechanism.

6. Thermodynamic Mechanism for Sexualization

According to the signaling network theory, sex corresponds to a special cellular signaling network. The sex distribution along the stimulus during the development of some species represents the thermodynamic distribution between two systems and the information integration between the genes and the environmental factors on sexualization (46–51).

Experimental scientists find that the sex formation of some species is mediated by certain factors, such as temperature (environment) determination (TSD), genetic determination (GSD), and simultaneous determination of genetics and temperature (47). Based on the thermodynamic model, the sex represents a special signal network and the sex formation represents the transformation of an unsexualized signaling network to a sexualized signaling network. The male-to-female ratio in the sex development of certain species can be described by the distribution laws of thermodynamics. Chromatin comprises the core structure of the sex-determining network. The information on environment (temperature) and genes are integrated into the signaling protein to determine the sex of a body. TSD and GSD represent two extreme cases of sex determination with numerous untypical cases. Even in species of TSD, numerous genes are related to sexualization.

An embryo cell has male and female sexualizing signaling activities. The male-to-female ratio of cell sexualization is determined by the ratio of the two types of signaling activities. Figure 9 shows the sexualization procedure of cells, and Figure 10 indicates several typical distribution curves. Figure 10Ab shows a typical sex distribution curve for certain species and a profile of the sexualizing

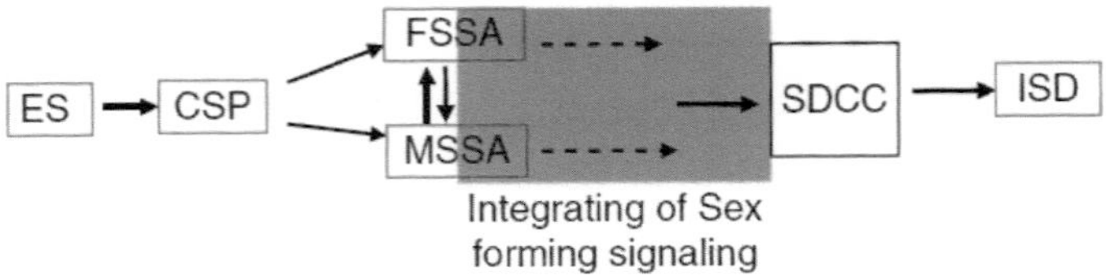

Figure 9. Signaling pathway of cell sexualization.

ES is the environment (temperature) signal, CSP is the cell-signaling protein that responds to environmental signals (physical or chemical) and transforms them into cellular protein signals, FSSA is the female sexualizing signal activity, MSSA is the male sexualizing signal activity, SDCC is the sex-determining chromatin conformation, and ISD is the irreversible sex development. FSSA and MSSA may be connected or separated. The two types of signaling activities are thermodynamically integrated during the chromatin formation and after the cell sexualization.

activity, and Figure 10Aa outlines the stability of sexualizing proteins. Figure 10Ab can be logically deduced from Figure 10Aa, which reflects that either the male or female sexualized signal activities are dependent on temperature. Figure 10B shows another case in which the typical profile of the stability curve (bell curve) of a sexualized signal can be observed within the range of experimental temperature. Figure 10C shows that the male-to-female ratio does not reach zero at extreme temperatures. There is a strict, logical relationship between the stability of the core structure and the sex distribution curve. As the core structure of sexualization is not altered, the change in the thermodynamic stability of the core structure is induced by a change in the protein quantities and species, which is in turn induced by the sexualized signaling activity. Therefore, the sexualizing protein and sex distribution, as shown in Figure 10, has an approximate relationship rather than a strict relationship. The relationship between sexualization and the thermodynamic properties of the signaling protein has been subjected to empirical investigation (50, 51).

The transition temperature of the distribution curve was previously referred to as pivotal temperature in the field of genesiology (46). The distribution curve shifts along the temperature as genetic changes occur among different stains (Figure 4 in reference 46), which strongly supports the proposed thermodynamic model. Such gene

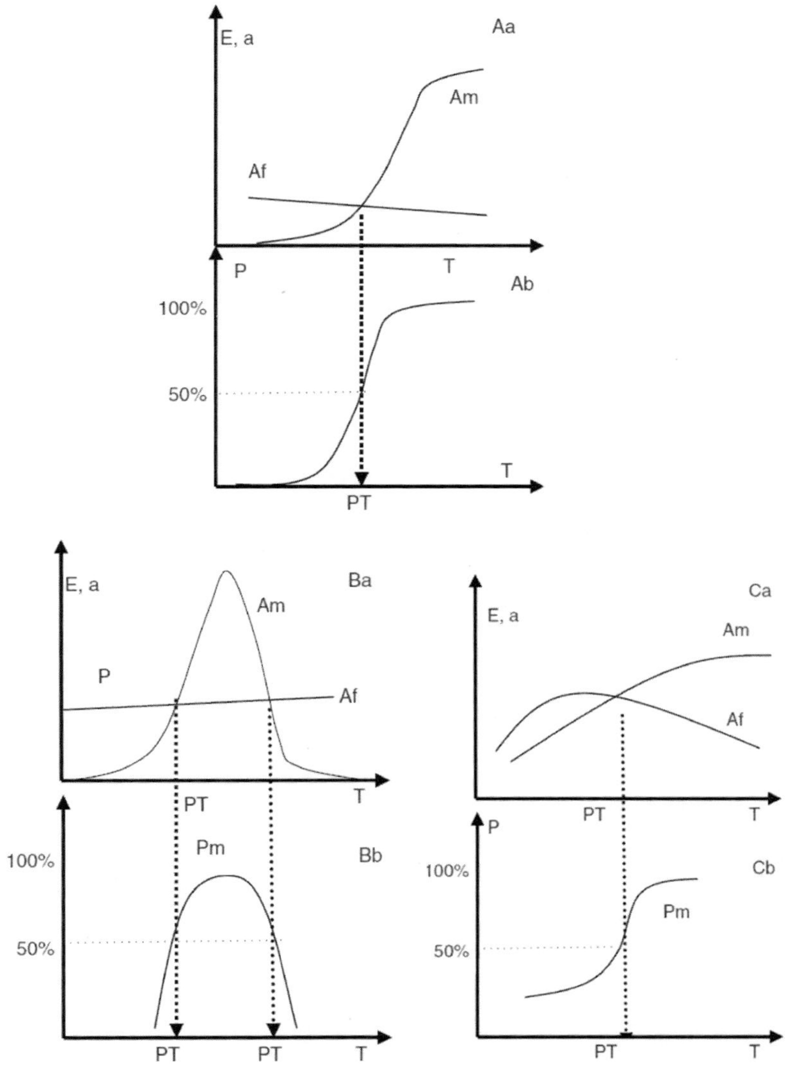

Figure 10. Several types of sex distribution.

T is the temperature, E is the stability and activity of the signaling protein, PT is the pivotal temperature, Am is the activity of male forming signaling, Af is the activity of female forming signaling, and Pm is the probability of male yielding. Figure A illustrates the typical distribution of two thermal states, Figure Bb shows the typical distribution curve with two PT, and Figure Cb shows the distribution curves of some species. As the activity of a signaling protein is related to protein stability, the difference between protein stability and activity is ignored in this figure.

mutation occurs in the signaling protein located along the pathway of signaling transduction, which begins from the environment to the chromatin. Such gene mutations do not occur in the core proteins of sexualization as general profile is not changed. The thermodynamic parameter of sex distribution must be carefully measured to explain the sex development processes, which has been insufficiently investigated.

7. Unstable Biosignal Network and Oscillation

The preceding discussion reveals the basic nature of the stable signaling network, which can remain in its present state forever or for a long time. Oscillation occurs between two unstable signaling networks (52, 53).

Equilibrium thermodynamic mechanism for the oscillation of signaling networks. The transformation between two networks can be thermodynamically and spontaneously processed when the Gibbs free energy levels of these networks reach the same level. The distribution between two networks can be precisely measured by the partition function concepts of a thermal system (19), such as the open and close states of a channel (54, 55). Figure 11 illustrates this principle, Figure 11a shows the basic principle of oscillation, Figure 11b shows the behavior of a channel that oscillates between open and close states, and Figure 11c shows the behaviors of many channels, such as those in the cell membrane.

Irreversible thermodynamic mechanism for the oscillation of signaling networks. An oscillation occurs between two protein signaling networks when a negative feedback regulation occurs between such networks. Figure 11d illustrates this principle. The spiking fashion of cerebellar purkinje neurons (56, 57) shows an ideal example of oscillation. The firing of neurons induces the closing of channels at their rest states and decreases their action potential as well. A mechanism to open the channel can produce another spike. Figure 11c shows the general profile of such oscillation.

Complex mechanism for the oscillation of cellular substances. The oscillation of a cellular substance, such as Ca^{++},

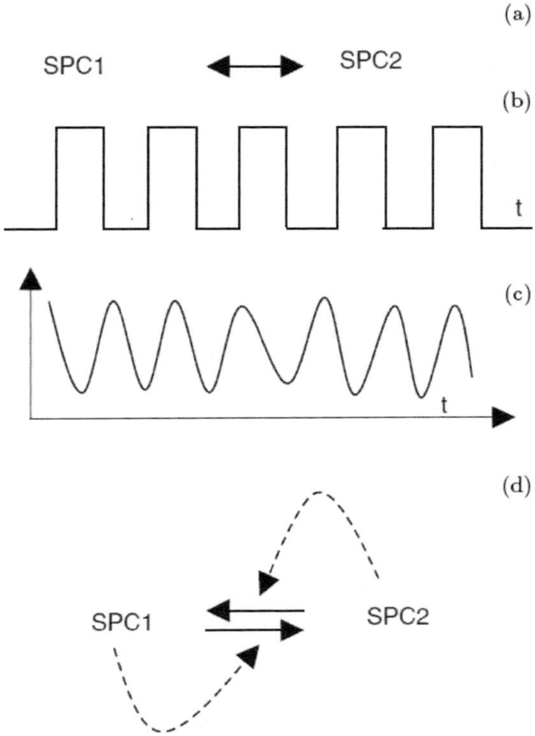

Figure 11. Oscillation of signaling network.

SPC1 and SPC2 are the signaling networks, and t is the time.

is more complex than that of signaling networks, given that such networks cannot be simply interpreted by any of the two basic mechanisms for oscillation (i.e., equilibrium and irreversible thermodynamics). Theoretically, these networks can be recognized as combinations of different mechanisms or signaling networks in which an experimental model is more appropriate than a theoretical model (58–60).

8. Conclusions and Perspectives

This paper outlines the differences between the currently adopted dynamic model of signaling network and the proposed thermodynamic model. The latter model has more advantages than the former model. First, while the logic independence of a signal network cannot be guaranteed within the dynamic model, such independence naturally exists within the thermodynamic model. Second, while the robustness of the dynamic model can hardly be understood (61), the different signaling networks of a cell can be freely combined and integrated in the proposed thermodynamic model, which improves the robustness of the signaling network. Third, a logical relationship exists between protein thermal motion (or thermodynamic properties) and biological function within the thermodynamic model, which cannot be observed within the dynamic model. Fourth, all types of information on genes and environmental factors can be integrated within the signaling protein in the thermodynamic model, but such information cannot be integrated within the dynamic model. Fifth, the cellular signaling network is a stable thermodynamic system (or structure) rather than a dynamic system. Sixth, the concept of attractors in the dynamic model is not compatible with the knowledge of molecular biology and biochemistry (62). Although the dynamic model of signaling network may have an advantage in explaining some complex phenomena of biology, it is not the basic principle of system biology.

The findings of this study are stated as follows. First, a logical relationship is established between system biology and thermodynamics because a signaling network represents one physiological and thermodynamic system. Second, the thermodynamic model can unify numerous concepts in experimental science and in theoretical science of system biology. Third, the thermodynamic model can explain various phenomena in system biology that can be found in many areas of experimental science. Fourth, the dynamic system is an unnecessary concept in the field of system biology. Fifth, a comprehensive understanding of system biology can be obtained based on the thermodynamic system. Overall, the proposed model provides a theoretical foundation for system biology based on thermodynamics, which can be applied in different fields of biology.

Plant grafting can also be explained by the proposed thermodynamic model, but such a model is not supported by sufficient empirical data.

The law of conventional thermodynamics does not support most previous studies on system biology, and the findings of previous studies cannot be quantitatively analyzed by such law. Therefore, a typical signaling network must be carefully and quantitatively analyzed in the future to expand the current understanding of biology.

References

1. Barabasi, A. L. & Oltvai, Z. N. Network biology: understanding the cell's functional organization. *Nat. Rev. Genet.* **5**, 101–113 (2004).
2. Oster, G. F., Perelson, A. S. & Katchalsky, A. Network thermodynamics: dynamic modelling of biophysical systems. *Q. Rev. Biophys.* **6**, 1–134 (1973).
3. Zhao, Q. Nonalgorithmicity and algorithmicity of protein science. Adv. Biosci. Biotech. **2**, 340–346 (2011).
4. Olayioye, M. A., Neve, R. M., Lane, H. A. & Hynes, N. E. The ErbB signaling network: receptor heterodimerization in development and cancer. *EMBO J.* **19**, 3159–3167 (2000).
5. Cheatham, B. & Kahn, C. R. Insulin action and the insulin signaling network. *Endocrine Reviews.* **16**, 117–142 (1995).
6. Schreiber, S. L. & Bernstein, B. E. Signaling network model of chromatin. *Cell,* **111**, 771–778 (2002).
7. Xiong, W. & Ferrell, J. E. A positive-feedback-based bistable 'memory module' that governs a cell fate decision. *Nature*, **426**, 460–465 (2003).
8. Kitano, H. Computational systems biology. *Nature*, **420**, 206–210. (2002).
9. Bhalla, U. S., Ram, P. T. & Iyengar, R. MAP kinase phosphatase as a locus of flexibility in a mitogen-activated protein kinase signaling network. *Science* **297**, 1018–23 (2002).
10. Ridgway, .D, Broderick, G. & Ellison, M. J. Accommodating space, time and randomness in network simulation. *Curr. Opin. Biotech.* **17**, 493–498 (2006).
11. Mikulecky, D. C. Network thermodynamics and complexity: a transition to relational systems theory. *Comput. Chem.* **25**, 369–91 (2001).
12. Qian, H. & Reluga, T. C. Nonequilibrium thermodynamics and nonlinear kinetics in a cellular signaling switch. *Phys. Rev. Lett.* **94**, 028101 (2005).
13. M. Hashemy and M. Monajjemi, Theoretical Study on Nanostructure of tRNA Using by Quantum Simulation Outlook. *Quantum Matter* 2, 214 (2013).

14. A. M. Ilyin, Computer Simulation of Nucleation and Formation of Armchair Carbon Nanotubes. *Quantum Matter* 2, 205 (2013).
15. Pankratov, E. L., & Bulaeva, E. A. (2013). Doping of materials during manufacture p–n-junctions and bipolar transistors. Analytical approaches to model technological approaches and ways of optimization of distributions of dopants. *Reviews in Theoretical Science*, **1**(1), 58–82.
16. Zhao, Q. A Molecular and Biophysical Model of the Biosignal. *Quantum Matter.* **2**, 9–16 (2013).
17. Zhao, Q. Protein thermodynamic structure. *IUBMB life*, **61**, 600–606 (2009).
18. Zhao, Q. Allodynamic Regulation of Protein Activity. *Quantum Matter*, **2**, 144–152 (2013).
19. Zhao, Q. Partition function of protein conformational state. *J. Comput. Theor. Nanosci.* **9**, 745–751 (2012).
20. Zhao, Q. Nature of Protein Dynamics and Thermodynamics. *Rev. Theor. Sci.* **1**, 83–101 (2013).
21. S. Amiri and J. Ali, Deform of biological human tissue using inserted force applied by optical tweezers generated by PANDA ring resonator. *Quantum Matter* **3**, 24 (2014).
22. I. S. Amiri, A. Nikoukar, J. Ali, and P. P. Yupapin, Ultra-short of pico and femtosecond soliton laser pulse using microring resonator for cancer cells treatment. *Quantum Matter* **1**, 159 (2012).
23. J. Herrou and S. Crosson, Function, structure and mechanism of bacterial photosensory LOV proteins. *Nat. Rev. Microbiol.* **9**, 713 (2011).
24. Wang, A., Zhang, Z. & Zhao, Q. In vivo measurement of protein functional changes. *Int. J. Biol. Sci.* **5**, 411–420. (2009).
25. Rinderknecht, H. Activation of pancreatic zymogens. *Dig. Dis. Sci.* **31**, 314–321 (1986).
26. Prusiner, S. B. Molecular biology of prion diseases. *Science*, **252**, 1515–1522 (1991).
27. He, Y. & Amasino, R. M. Role of chromatin modification in flowering-time control. *Trends Plant Sci.* **10**, 30–35 (2005).
28. Burn, J. E, Bagnall, D. J, Metzger, J. D, Dennis E. S. & Peacock, W. J. DNA methylation, vernalization, and the initiation of flowering. *Proc. Natl. Acad. Sci. USA*, **90**, 287–291 (1993).
29. Finnegan, E. J., Peacock, W. J. & Dennis, E. S. DNA methylation, a key regulator of plant development and other processes. *Curr. Opin. Genet. Dev.* **10**, 217–223 (2000).
30. Xu, R., Nelson, C. M., Muschler, J. L., Veiseh, M., Vonderhaar, B. K. & Bissell, M. J. Sustained activation of STAT5 is essential for chromatin

remodeling and maintenance of mammary-specific function. *J. Cell Biol.* **184**, 57–66 (2009).
31. Wood, L. D. *et al.* The genomic landscapes of human breast and colorectal cancers. Science, **318**, 1108–1113 (2007).
32. Withers, D. J. & White, M. Perspective: the insulin signaling system — a common link in the pathogenesis of type 2 diabetes. *Endocrinology*, 141, 1917–1921 (2000).
33. Burdon, T., Smith, A. & Savatier, P. Signalling, cell cycle and pluripotency in embryonic stem cells. *Trends cell biol.* **12**, 432–438 (2002).
34. Kreeger, P. K. & Lauffenburger, D. A. Cancer systems biology: a network modeling perspective. *Carcinogenesis*, **31**, 2–8. (2010).
35. Michaels, S. D. & Amasino, R. M. Memories of winter: vernalization and the competence to flower. *Plant Cell Environ.* **23**, 1145–1153 (2008).
36. Amasino, R. M. Vernalization and flowering time. *Curr. Opin. Biotechnol.* **16**, 154–158 (2005).
37. Shelford, V. E. Some concepts of bioecology. *Ecology*, **12**, 455–467 (1931).
38. Streck, N. A., Weiss, A. & Baenziger, P. S. A generalized vernalization response function for winter wheat. *Agron. J.* **95**, 155–159 (2003).
39. Streck, N. A. A temperature response function for modeling leaf growth and development of the African violet (Saintpaulia ionantha Wendl.). *Ciência Rural,* **34**, 55–62 (2004).
40. Díaz, A., Zikhali, M., Turner, A. S., Isaac, P. & Laurie, D. A. Copy Number Variation Affecting the Photoperiod-B1 and Vernalization-A1 Genes Is Associated with Altered Flowering Time in Wheat (Triticum aestivum). *PLoS ONE*, **7**, e33234. (2012).
41. Ferrell, J. E. How responses get more switch-like as you move down a protein kinase cascade. *Trends Biochem Sci.* **22**, 288–9 (1997).
42. Evans, L. & Dunstone, R. L. Some physiological aspects of evolution in wheat. *Aust J Biol Sci*, **23**, 725–742 (1970).
43. Bissell, M. J., Hall, H. G. & Parry, G. How does the extracellular matrix direct gene expression? *J. Theor. Biol.* **99**, 31–68 (1982).
44. Bissell, M. J. & Hines, W. C. Why do we get so few tumors? A proposed role of the Microenvironment in restraining cancer progression. *Nat. Med.* **17**, 320–9 (2011).
45. Park, C. C., Zhang, H., Pallavicini, M., Gray, J. W., Baehner, F., Park, C. J. & Bissell, M. J. Integrin inhibitory antibody induces apoptosis of breast cancer cells, inhibits growth, and distinguishes malignant from normal phenotype in three dimensional cultures and in vivo. *Cancer Res.* **66**, 1526–35 (2006).

46. Mitchell, N. J., Nelson, N. J., Cree, A., Pledger, S., Keall, S. N. & Daugherty, C. H. Support for a rare pattern of temperature-dependent sex determination in archaic reptiles: evidence from two species of tuatara (Sphenodon). *Front Zool.* **3**, 1–12 (2006).
47. Bull, J. J. Sex determination in reptiles. *Quart. Rev. Biol.* **55**, 3–21 (1980).
48. Mrosovsky, N. & Pieau, C. Transitional range of temperature, pivotal temperatures and thermosensitive stages for sex determination in reptiles. *Amphibia-Reptilia.* **12**, 169–179 (1991).
49. Deeming, D. C. & Ferguson, M. W. J. Environmental regulation of sex determination in reptiles. *Phil. Trans. R. Soc. Lond.* **322**, 19–39 (1988).
50. Crews, D. Temperature-dependent sex determination: The interplay of steroid hormones and temperature. *Zool Sci.* **13**, 1–13 (1996).
51. Devlin, R. H. & Nagahama, Y. Sex determination and sex differentiation in fish: an overview of genetic, physiological, and environmental influences. *Aquaculture,* **208**, 191–364 (2002).
52. Wijnen, H. & Young, M. W. Interplay of circadian clocks and metabolic rhythms. *Annu. Rev. Genet.* **40**, 409–48. (2006).
53. Ferrell, J. E., Tsai, T. Y. & Yang Q. Modeling the cell cycle: why do certain circuits oscillate? *Cell,* **144**, 874–85 (2011).
54. Hammond, C. *Cellular And Molecular Neurophysiology,* (3rd ed. Academy press, London, 2008).
55. White, J. A., Rubinstein, J.T. & Kay, A.R. Channel noise in neurons. *Trends Neurosci.* **23**, 131–7 (2000).
56. Womack, M. D. & Khodakhah, K. Somatic and Dendritic Small-Conductance Calcium-Activated Potassium Channels Regulate the Output of Cerebellar Purkinje Neurons. *J. Neurosci.* **23**, 2600–2607 (2003).
57. Womack, M.D. & Khodakhah, K. Dendritic Control of Spontaneous Bursting in Cerebellar Purkinje Cells. *J Neurosci.* 24, 3511–3521 (2004).
58. Sneyd. J., Keizer, J. & Sanderson, M. J. Mechanisms of calcium oscillations and waves: a quantitative analysis. *FASEB J.* **9**, 1463–1472 (1995).
59. Y. S. Shin, F. Remacle, R. Fan, K. Hwang, W. Wei, H. Ahmad, R.D. Levine, J. R. Heath, Protein signaling networks from single cell fluctuations and information theory profiling. *Biophys. J.* **100**, 2378–86 (2011).
60. Sachs, K., Perez, O. & Nolan, G. P. Causal protein-signaling networks derived from multiparameter single-cell data. *Science,* **308**, 523–529 (2005).

61. Kitano, H. Biological robustness. *Nat. Rev. Genet.* **11**, 826–37 (2004).
62. Takens, F. Detecting strange attractors in turbulence. Springer Lecture Notes in Mathematics. **898**, 366–381 (1981).

Chapter 26

Biological Development

Summary: The general principles of biological development are discussed on the basis of the irreversible thermodynamic and signaling network theory. It is shown that the bio-informational flow from a gene to its biological function is a process of informational integration, rather than the translation process. The integration process cannot be described by a unified program (or model), thus the system relation between genes and biological function has been proposed.

A biological organism is developed from a fertilized egg or oosphere. When the egg is fertilized, it has an ability to develop into many types of stem cells, can be further developed into the diversified organ cells of body and finally a developed organism is formed (1). Developmental biology has attracted much attention from many scientists, and with most scientists working in the field of the experimental science, a lot of empirical results have been obtained. Many empirical views about it has been proposed, but there has been no satisfactory theory until now (2). In view of irreversible thermodynamics, biological development is a typical process for the origin of natural order (from nothing to everything). The oosphere can be recognized as nothing and the diversified species of cells of an organism represent everything.

It is necessary to differentiate growth from development. From the perspective of general biology, growth refers to an increase of body weight (or a cell), or the quantity of one's biological function; while development refers to the emergence of new functions (full functioning of the body or maturity). From the perspective of cell biology, growth describes a change of cell numbers of same species, while development represents the cell transformation from one species (type) of cell to another species of cell. From the perspective of the bio-signal network theory, growth represents the quantitative change of the signaling activity of a signal network, and development refers to the emergence of a new type of signal network.

At present, most experimental scientists have focused on revealing the detailed pathways of the signaling transduction pathway involved in biological development. Unfortunately, there are still not enough references on this topic. In this chapter, we will discuss the general properties of it on the basis of irreversible thermodynamics. Of course, what is discussed here cannot describe all the traits of developmental biology.

1. Biological Development is Logically Related to the Properties of Molecular Machines

When we study molecular biology or biochemistry, we learned enzymes or proteins of microorganisms are the same as that of a human or animal; enzymes isolated from different species have the same activity, similar protein sequences, and even similar structure. For any detailed property, the protein isolated from a microorganism is not inferior to that from a human. Furthermore, we have leant that the animals, or more precisely for the human and Apo, shared similar biological machines (near 100%) at molecular level; if not all, at least most molecular machines of a human can be replaced by a monkey's without changing its ability. For example, the insulin of animals, such as bovines and pigs, work well in the human body and are commonly used as medicine to treat diabetes. Thus, the biological development is not logically determined by the properties of molecular machines.

One view is that biological development results from the cooperation of many molecular machines *in vivo* and that biological development consists the behavior of whole body, rather than any single molecular machine in the body. Although this sounds plausible, this theory obviously ignores essential differences between properties of molecular machines among different biological species.

A more balanced argument is that biological development is related to the properties of molecular machines, are not through determined by these properties. We can understand it by several aspects of proteins.

Firstly, although proteins isolated from different biological species have similar properties and functions, they are differentiated by some properties, particularly in their binding ability to other protein species. For example, although the bovine bFGF (Basic fibroblast growth factor) has similar functions as the human bFGF, the bFGF of human is much more active than the bovine bFGF.

Secondly, if a protein complex has two different partners, the formation of this protein complex is determined by both partners. The binding between different protein subunits is determined by the properties of both subunits.

Thirdly, From the perspective of genetics, humans have more species of proteins than that of lower organisms.

Fourthly, there is a vital relationship among all types of signaling activities of organism because they take up a lot of space, consume energy and bio-material, which is essential for other signaling activities (biological functioning).

In summary, we conclude that molecular machines act as fundamental materials of biological development, but it does not logically determine biological development alone.

2. The Irreversible Process of Biological Development from the Perspective of Thermodynamics

According to the principles of irreversible thermodynamics, we have classified the growth of an organism is the equilibrium process (reversal process) and biological development as an irreversible process. Facts that supported this augment were summarized as follows:

1) Currently, we believe that the change of body weight is a reversible process of thermodynamics and can be described by the laws of equilibrium thermodynamics.

2) Abilities and skills, such as riding a bicycle and walking, are retained even if we have not used them for a longtime. It is a nature of the irreversible process of thermodynamics.
3) Wheat vernalization can be reversed if it is not finished; it is reversible process. A plant that has been entirely vernalized cannot reverse into unvernalized state; it is irreversible process.

From the perspective of thermodynamics, the biological development provides the boundary of equilibrium thermodynamics of organism. The growth of the body provides materials and energy which is essential for the biological development. Biological development in turn controls the growth of the body and represents the transformation of the signaling network which provides boundary condition for growth; while growth represents represent signaling activity of signal network at fundamental level. Biological development is a qualitative change of a signaling network and growth is a quantitative change of signaling network. Biological development is irreversible process and growth is reversal process from the perspective of thermodynamics. There is a typical logic cycle between them.

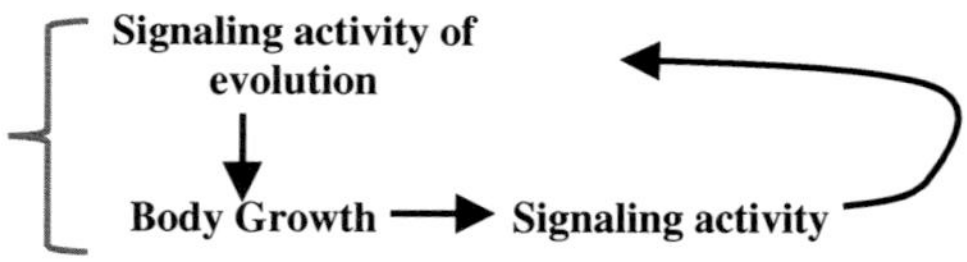

This is a typical structure of the system.

At the time of development, the manner of biological regulation changes. Growth cannot alter the regulation patterns.

3. An Irreversible Process of Gene Expression and Cell Speciation

Development is related to gene expression, and some new types of signal networks, such as cell speciation, are accompanied by the expression of a new protein. Thus, the gene regulation theory is significant in our understanding of biological development. The operon theory, which involves the operon and has been widely discovered in microorganisms, has been well documented in textbooks and dominates the thought of many scientists (3, 144). Within this model, there is inducer for an operon and the gene expression of this operon can be induced by this inducer. When the inducer is added to a culture medium, the gene will be expressed immediately, and when inducer is removed from culture medium, the gene expression will end as well. According to the operon theory, gene expression is reversal process and its traits can be precisely described by principles of equilibrium thermodynamics.

However, no operon has ever been discovered in a eukaryote and the biological development of eukaryote is processed irreversibly. The best known example of biological development is the vernalization of winter wheat or other plants (5, 6). Without verbalization, a wheat can grow, but cannot reproduce. Temperature is the environmental factor for plant vernalization, but it is not the inducer of an operon for plant vernalization. An plant vernalization is also a specific case for cell speciation, in which unvernalized cell is transformed into a vernalized cell and can further develops into the whole body of a plant.

A stable system of thermodynamics, or the signal network of biology, should comprise of all types of signaling activities. When vernalization is completed, the effect of temperature on plant vernalization should be disappear. We have drawn Figure 1 to illustrate about features of cell speciation according to principles of the signal network theory (as shown in Chapter 25 of this book).

The core structure of the signal network of cell speciation is chromatin conformation. The DNA methylation plays important role in gene expression and the epigenetics of biological traits. However, these will not be covered in detail in this book, for further references, reader can refer Ref. 7 and 8.

4. The Giant Space of Physiological States for Biological Development

According to the theory of the origin of natural order (or irreversible thermodynamics), a giant independent space (IS) can be generated

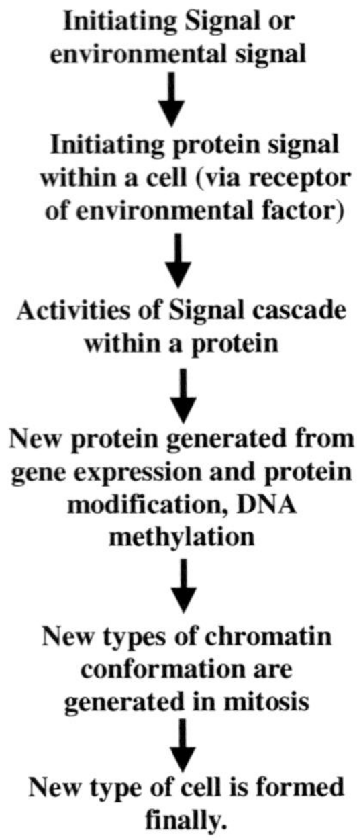

Figure 1. The procedure of cell speciation.

The process of cell speciation is shown in this Figure. (Note: not all aspects of signaling activities are expressed in this Figure, we will only be studying particular parts of it)

from the motions at fundamental level and which properties cannot be explained fully by properties of material at fundamental levels. In the field of developmental biology, the information at fundamental level is DNA sequence or genomes. DNA is giant molecules and it contains all types of information needed for biological development. However, the information of DNA should be firstly transformed into signaling activity (or physiological activity) and it can play a role in biological development. In one way, the DNA sequence was translated into proteins sequence, information of internal motion within a protein, finally, information of protein thermodynamic structure, information of protein regulation, protein interaction, and so on. All these machines or organs represent giant independent space (IS). Here we summarized some IS as following Figure.

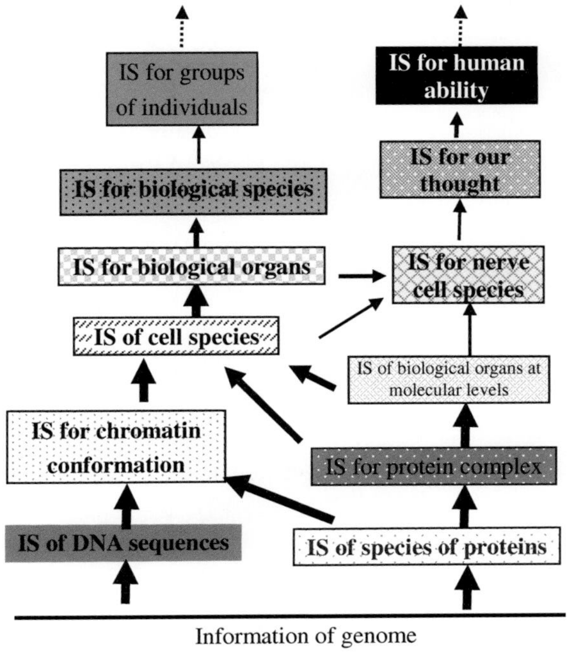

Figure 2. The existing IS on diversified levels of biology or developmental biology.

This Figure has summarized some types of IS which are at diversified levels of biology. (Note: not all types of IS in biology have been pictured in this Figure as some types of IS are out of the scope of developmental biology)

All the types of IS represented in Figure 2 are well known in biology. For example, the IS of chromatin conformation has been recently revealed; DNA loop and the methylation of DNA play a key role in the conformation of chromatin, which controls the expression of genes and epigenetics (7).

5. The Biological Development is the Process of Information (Signal) Integration

All types of information on genes or on environment, are integrated at diversified levels of biological organs, such as protein. The interaction among different genes represents an integration mechanism of bio-information of gene at different levels. In this case, information on genes is firstly transformed into information on peptide sequences, and then into thermodynamic information (or thermodynamic parameter, internal motion of components of a protein) on residues, and integrated at one level of signaling activity of proteins. The integration of these types of information in process of protein folding results in the formation of protein three-dimensional structure or a protein thermodynamic structure; the integration of information during the protein active conformation is the protein regulation process; and so on.

6. The System Relation Between Biological Function and Gene

In our study of genetics, we have learnt that there is no absolute definition for the concept of gene (9, 10). If we assume that a gene is the basic unit for a crossover, it must be a base which cannot represent a biological trait or function; if we assume that a gene has enough information for a biological function (or trait), it cannot act as basic unit of a crossover. Thus, we must find the definition of a gene or its meanings in a given context, which have resulted in several revision of the gene concept in history. In view of the system theory, it is a natural phenomenon.

Considering that a gene or genome represents a system and biological functions are also systems, their relationship can be only described by system logic. By applying principles of system logic, the following conclusions can be obtained.

1) **For any biological functions and its related genes, the linear relationship between gene and biological functions can be isolated within a model and validated in a given condition.**
2) **Numerous models of such models can be isolated.**
3) **These models, in which the linear relation between the gene and biological function are validated, cannot be annexed into a unified model for they are validated under different cases.**

For example, in vivo and normal condition, the cytochrome c is one of the many proteins of the electron transport chain in mitochondria and its activity is necessary for the survival for a cell. However, it is also a protein involved in cell apoptosis. These two functions are not compatible with each other. The key reason is that biological information can be integrated via different ways and one model can only describe the properties of one fashion of the integration mechanism.

7. The Inheritance of Biological Trait Within IS

As we have discussed, there are many ISs in biology at diversified levels of biology; within such ISs, properties of the biological traits or (characteristics) at advanced levels of biology cannot be fully explained by the properties of materials (or laws) at fundamental levels. So inheritance of biological traits can be observed at any IS of biology. Examples are prions (its protein conformation can be propagated forever), cell species (remains in its state for ever), or even our culture which is inherited through our daily activity (11). Cancer is also an inherited trait, but the gene mutation of a human genome can enhance the emergence of the cancer cells or increase sensitivity of an individual to cancer (12, 13). Some diseases, such as diabetes, may be inherited via cell division (asexual reproduction, but bot sexual reproduction).

The criteria for the judgment of biological trait with inheritance is whether the trait is stable in view of thermodynamics. The nerve cell and the liver cell have the same genetic background, and can remain their nature in a same medium but cannot transform with each other freely, so the cell species is an acquired inheritance.

Within our model, the inheritance of biological traits is common phenomenon of biology and can be transmitted via way of asexual reproduction. It is compatible with the gene theory completely.

The inheritance of an acquired trait may be considered a specific case for the inheritance of a biological trait. Our theory indicates that acquired traits can be classified into two groups. The first is same as the inheritance of biological trait, which can be transmitted via the way of asexual reproduction. The second is gene modification, for example the transformation of a silence gene to active gene, which can be transmitted via sexual reproduction.

The term epigenetics has been well utilized to diverse fields of biology. My opinion is that the inheritance of biological traits within an IS can account for all phenomena of epigenetics.

8. The Phasic Development of Plants

It has been long proposed that the development of plants shows different physiological stages (14). The most typical example is plant vernalization. This example supports our theory. Within our theory, many types of physiological states or signaling networks are passed over in the development of a plant or an animal. Each physiological states represents one physiological stages. But in agricultural science, only a few physiological stages can be isolated and demonstrated, most biosignal networks cannot be isolated empirically in agricultural science. Here it is conveniently to name a new concept: developmental space. It represents a set of all possible physiological states (Fig. 3) within it a body can be developed on the basis of a genetic background.

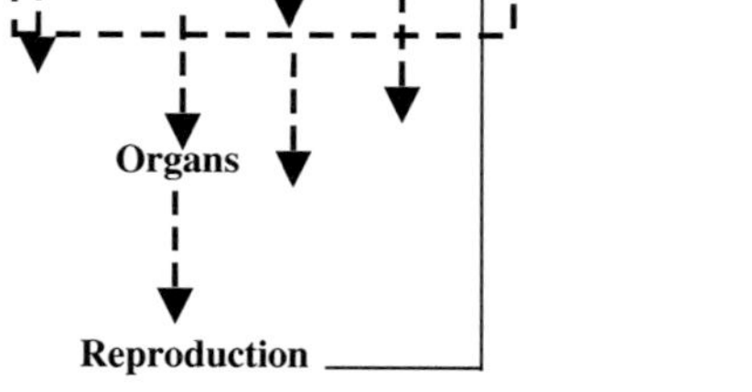

Figure 3. The diagram for biological development and physiological stage.

The fertilized egg undergoes many steps of cell speciation and organogenesis, and finally the individual is developed. In this process, diferent physiological stages can be produced. A physiological stage may include many different cells or signal networks.

9. The Signal Redundancy for Biological Development

Information or signal redundancy is a well-established theory in the field of developmental and molecular biology (15). It states that the genomes of any species have more information than the amount required for biological development or biological regulation.

For example, in cancer biology, numerous suppressed and cancerous genes are isolated. It indicates that a biological trait can be developed via different ways or through mechanisms at molecular level.

From the perspective of thermodynamics, it is the most natural phenomenon. A biological trait represents a thermal system, thus suppressed and activated genes are its constituents, and which can be structured differently (or related differentially). The redundancy of a signal or information provided extra stability for the signal network or biological function.

10. The Integration of Environmental Signal and Genetic Signal

The biological development and biological function is either under control of genetic or the environment (16). In a sense, the development of all biological organs or functions are control by the environment. However, in the cases as in the development of sex, the genetic background of our body has provided all the information (including information to provide a suitable environment) required, and it is controlled by genetics. In fact, there are a lot of reports on abnormal sexual development, which could have been induced by a genetic (gene mutation) or environmental change (or induced by toxicants). In other words, from the perspective of molecular biology, the development of all biological functions or organs are control by the environment; from the perspective of genetics, the development of some biological functions may be genetically control under normal conditions.

In fact, environmental and genetic control are two extreme cases in biological development. In most cases, there is signal integration between environmental signals and genetic signals, and integration mechanism can be described by partition functions of a system as we have discussed in previous chapters.

11. The Abnormal Physiological States of an Organism

A number of physiological states or signal networks can be generated from the same genetic background. Some of it can be generated normally during biological development in a certain stage, while some may be generated only under specific conditions, and others may appear in the condition of a disease.

Not all possible physiological states generated in one genetic background are healthy for some are indeed harmful to our health and may be an indicator for disease. But there is little research about that and we will not discuss it here in detail.

References

1. Wobus, A. M., & Boheler, K. R. (2005). Embryonic stem cells: prospects for developmental biology and cell therapy. *Physiological reviews*, 85(2), 635–678.
2. Thomson, J. A., Itskovitz-Eldor, J., Shapiro, S. S., Waknitz, M. A., Swiergiel, J. J., Marshall, V. S., & Jones, J. M. (1998). Embryonic stem cell lines derived from human blastocysts. *science*, 282(5391), 1145–1147.
3. Alberts B, Johnson A, Lewis J, *et al.* (2002) Molecular Biology of the Cell. 4th edition. New York: Garland Science;. Universal Mechanisms of Animal Development.: http:// www.ncbi.nlm.nih.gov/ books/ NBK26825/
4. Nelson, D. L., Lehninger, A. L., & Cox, M. M. (2008). Lehninger principles of biochemistry. Macmillan.
5. Sheldon, C. C., Finnegan, E. J., Rouse, D. T., Tadege, M., Bagnall, D. J., Helliwell, C. A., ... & Dennis, E. S. (2000). The control of flowering by vernalization. *Current opinion in plant biology*, 3(5), 418–422.
6. Sung, S., & Amasino, R. M. (2005). Remembering winter: toward a molecular understanding of vernalization. *Annu. Rev. Plant Biol.*, 56, 491–508.
7. Mirkovitch, J., Mirault, M. E., & Laemmli, U. K. (1984). Organization of the higher-order chromatin loop: specific DNA attachment sites on nuclear scaffold. *Cell*, 39(1), 223–232.

8. Finnegan, E. J., Genger, R. K., Peacock, W. J., & Dennis, E. S. (1998). DNA methylation in plants. *Annual review of plant biology*, 49(1), 223–247.
9. Portin, P. (2002) Historical development of the concept of the gene. *Journal of Medicine and Philosophy*, 27, 257–86.
10. Gerstein, M. B., Bruce, C., Rozowsky, J. S., Zheng, D., Du, J., Korbel, J. O., ... & Snyder, M. (2007). What is a gene, post-ENCODE? History and updated definition. *Genome research*, 17(6), 669–681.
11. Hsiao, K., & Prusiner, S. B. (1990). Inherited human prion diseases. *Neurology*, **40**(12), 1820–1820.
12. Jones, P. A., & Laird, P. W. (1999). Cancer-epigenetics comes of age. *Nature genetics*, 21(2), 163–167.
13. Baylin, S. B., & Herman, J. G. (2000). DNA hypermethylation in tumorigenesis: epigenetics joins genetics. *Trends in Genetics*, 16(4), 168–174.
14. Angus, J. F., Cunningham, R. B., Moncur, M. W., & Mackenzie, D. H. (1980). Phasic development in field crops I. Thermal response in the seedling phase. *Field Crops Research*, 3, 365–378.
15. Nowak, M. A., Boerlijst, M. C., Cooke, J., & Smith, J. M. (1997). Biological development of genetic redundancy. *Nature*, 388(6638), 167–171.
16. Jaenisch, R., & Bird, A. (2003). Epigenetic regulation of gene expression: how the genome integrates intrinsic and environmental signals. *Nature genetics*, 33, 245–254.

Chapter 27

Hybridization and Hybrid Vigor

Summary: Hybridization is a common process in biological reproduction. The change in the genetic background of an organism has a great impact on all phenotypes and biological functions of the organism. Among all the alterations, which are induced by a change in genetics, hybrid vigor is a well-established phenomenon. The working mechanism of hybridization is discussed on the basis of biological development and the signal network theory. It shows that many mechanisms have made a contribution to the production of heterosis and the main mechanism is the loose control of a physiological activities of organism.

Hybrid vigor, or heterosis is a well-known process in the field of biology and has attracted much attention in the history of biology (1–2). As a fact, it is well accepted, but there is currently no consensus about working mechanism behind it. Due to historical reasons, most scientists study it by applying principles of genetics. Currently, there are two different augments pertaining its working mechanism.

The first or the prevailing argument about its working mechanism was established on the basis of genetics. Although there are many of revisions about it, the key idea behind this hypothesis is that there are many alleles for a gene in a species; the hybrid offspring carrying more alleles can gain an advantage from the dominating gene or the interaction among these alleles (2–3).

The second argument was proposed by Ivan Vladimirovich Michurin. The key idea of this model is that an organism can develop different inheritable traits under different environments (now termed as epigenetics), with each environment pertaining one fashion of a physiological system; the blending of these different physiological systems by hybridization can change the phenotype of the hybrid offspring, which may result in hybrid vigor (4–9).

In this chapter, we will present a third argument about hybrid vigor. The theoretical foundation for our theory is the signaling network theory for physiological systems and the irreversible theory for biological development. It states that chromosomes (or background of genetics) are the main information with regards to inheritance and any alteration of it by hybridization has great impact on the development, fashion of regulation of physiological system, and all other traits of an organism. Hybrid vigor can be observed for some types of hybridization, but not for all hybrids. It gives a satisfactory explanation to first and second argument about hybrid vigor.

The change of the genetic background of an organism (or genomes) has great impact on all types of biological functioning and the trait of the organism at diversified levels of biology. If we wish to reveal the working mechanism of hybrid vigor, we should give an explanation for all matters that occurs at diversified levels of biology. In view of experimental science, it is impossible to do so at present. Thus, our discussion is limited to the parameters of finding a way to understand the working mechanism behind hybridization on the basis of the basic theory of biology.

The asexual hybridization of a plant represents a specific case of hybridization: the hybridization of physiological or signaling activities between the stock and the scion. Theoretically, the change observed in the hybridization of genetic material or gene recombination is also processed via the recombination of physiological activities between two species. Compared with asexual hybridization, which can only modify physiological state of the scion smoothly, sexual or genetic hybridization can greatly alter all phenotypes or physiological states of an organism, with the retention of the fundamental principles. Thus, we can discuss these general principles of them without differentiating them in some cases.

1. The General Features of Hybridization

The change of genetic background of an organism via hybridization has a great impact on the physiology of the organism. Two of these impacts, which are of great importance in our discussion about the working mechanism for hybrid vigor, are the expansion of developmental space of the hybrid and the quantitative change of the property of a signal network, or biological functioning.

Via hybridization, the hybrid has more species of the gene or has much more information on genetics than that of their parents; therefore, it provides more valuable genes for biological development. In terms of developmental biology, the hybrid has a large developmental space which allows it to show more phenotypes under the different conditions. This argument has been well established in the field of traditional genetics (2, 3).

A change of the genetic background will result in numerous alterations of phenotypes of an organism, and here we will cite and summarize several typical results in Figure 1 and discuss as follow:

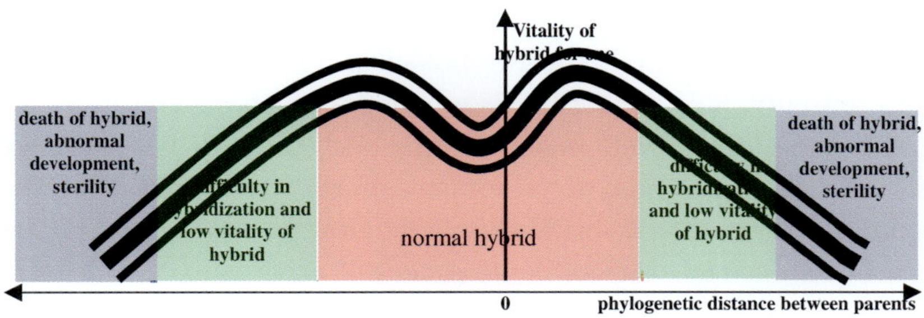

Figure 1. The relations of a number of biological phenomena observed in hybrid

The general relations of the hybrid vigor and a variety of related phenomenon are shown in this Figure (This Figure is the summarization of numerous results of hybridization and it describes general features of hybridization; it cannot be understood in a broad sense).

Some concepts and results about the hybrid vigor has been summarized in Figure 1 and we will briefly discuss its general mechanism and others here.

If the parents of the hybrid come from the same species and share similar genetics background (red area of Figure 1), the hybridization could be processed well and hybrid show diversified phenotypes. In this case, the dominating and valuable genes can be enriched in hybrid and which will result in hybrid vigor. This mechanism has been well documented previously.

If parents of hybrid come from different species and share little similarity in terms of their genetic background (dark area of Figure 1), the hybrid will show difficulty during development and low vitality due to the incompatibility between two physiological systems of its parents.

If the phylogenetic distance between the parents is larger, the hybridization cannot be processed naturally by sexual reproduction. But at cell level, two cells can be hybridized, but the fused cell will have no ability to develop because of physiological incompatibility. In agriculture, different plants can be grafted into one living body and tis is called distant hybridization.

Theoretically, the developmental space of a hybrid is changed during hybridization and new types of physiological states can be generated within the body of the hybrid. But at present, it is difficult to isolate new physiological state (signal network) of an organism experimentally. However, several biological phenomena indicate that occur actually. Firstly, sterility is a common phenomenon in hybrid. In this case, the development of the body cannot be processed. Secondly, the abnormal development of the hybrid is commonly found. Thirdly, some hybrids show low vitality. All these represent new, but abnormal state of development of an organism.

2. The Direction of Biological Development can be Changed

If the genetic background of an organism is changed, the development space of it will be altered too, which may result in the generation of many different phenotypes of the hybrid. One of them is the plasticity of biological development. Here we will discuss several typical cases about it.

First, if there has been little interference among the physiological states of two parents of hybrid, the development space of hybrid will

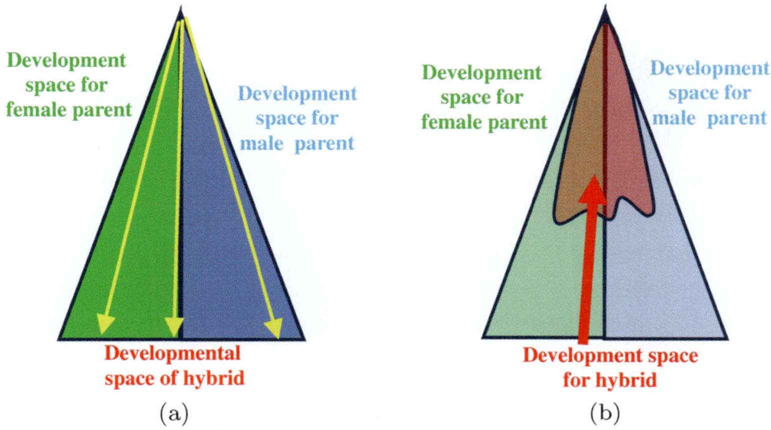

Figure 2. Developmental space and direction of development

In Figure 2a, the developmental space of the hybrid comprises of both developmental space of its parents and the direction of biological development of the hybrid can be changed (yellow). In Figure 2b, the developmental space of hybrid is very small (red area).

be the sum of that of their parents (Figure 2a). Hybrid vigor could be observed in this type of hybridization. The biological development of this hybrid will show a high plasticity than that of its parents and the hybrid will have a high survival rate under some living conditions. Examples of these types of hybrids are the hybridization of winter wheat and spring wheat.

Second, if there is a strong interference (or cooperation) among the physiological states of the hybrid's parents, the development space of hybrid will be reduced as compared with their parent (Figure 2b). The extreme case is that the system development of hybrid cannot be processed and hybrid dies at different stages of the development.

Third, if there is little cooperation between physiological states of hybrid's parent, new types of physiological states (function or organ) may be produced. The hybrid can then act new biological species. In this case, the direction of the biological development of an organism is also changed.

3. The Cooperation and Interference of the Different Physiological States

When two physiological states within one cell are forced to be expressed, they may remain independence in some cases. In most cases, however, the interference or cooperation between them or their components will occur.

Let us discuss this scenario with an example of a protein with two subunits. If the parents of hybrid are homozygous and their protein have two same subunits, but the proteins of the hybrid are different (see Figure 3), three phenotypes of hybrid will be produced.

In Figure 3, the recombination of the two subunits of a protein in the hybrid is shown. When two different subunits of a protein are combined, several results could be produced. If subunits of the

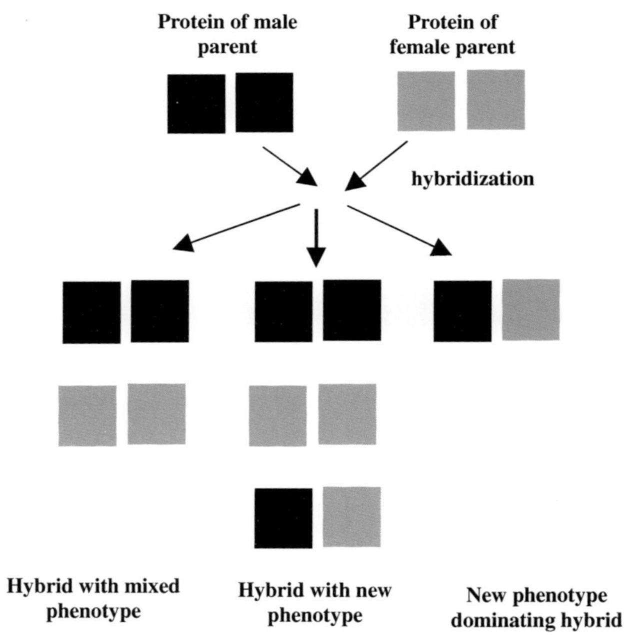

Figure 3. phenotype of hybrid with protein subunit reorganization

In this Figure, the two parents are homozygotes and have a protein with the same subunits, with a difference in the sequence of protein in the maternal and paternal parents. The three phenotypes of a hybrid could be produced.

protein coming from different parents cannot bind to each other, the hybrid (hybrid with mixed phenotype in Figure 3) will have two different protein species. If the subunits coming from their parents has similar binding ability, the hybrid will show a new phenotype (hybrid with new phenotype in Figure 3). If the subunits coming from different parents have a particular binding ability, only the hybrid protein can be generated (new phenotype dominating hybrid in Figure 3).

For a specific hybrid, these phenotypes of the protein subunit recombination may not show hybrid vigor, but for numerous types of hybrid, we can ascertain that some types of recombination show hybrid vigor.

If the recombination of protein constituents occurs in a regulatable protein complex in one pathway of signaling transduction, the case will be changed, and the fashion of the physiological regulation of hybrid and the gene expression will also be changed.

4. The Change of the Physiological Regulation of a Hybrid as a Working Mechanism for Hybrid Vigor

As we have discussed in Figure 3, the recombination of protein subunits within a protein can alter the protein properties and its constituents. If this type of recombination occurs in a signaling protein, such as a protein complex or chromosome (a type of protein-DNA complex), the regulatable fashion of physiology of an organism will be altered. This change may be the major mechanism among many mechanisms which contributes to hybrid vigor.

Numerous evidences about biological regulation have been presented. For example, the change of gene expression fashion of a hybrid has been observed, and it results from a change of biological regulation (10, 11). But the detailed molecular mechanism behind this phenomenon has not been fully exposed.

When a wild plant and a domesticated plant are hybridized, the traits of the wild plant are always dominating in F1 offspring. Thus we concluded that physiological activities of the wild plant are under strict control as compared to that of domestic and hybrid plants. We further conclude that the working mechanisms of both hybrid vigor and the domestic plant (or domestic animal) are under loose control.

Next we will show that this hypothesis can give a reasonable explanation for most of the observations in the field of agriculture and hybridization. It is not the sole mechanism for hybrid vigor, but it contributes much to most aspects of hybrid vigor.

5. The Feature of the Wild and the Domestic Organism

Wild plants that live in barren land must endure many unfavorable environmental factors or conditions, which are far away from the optimal condition for plants to grow. In order to survive in these conditions, these plant are capable of adapting different living conditions.

However, the domestic plants grow in fertile soil with enough water and fertilizers. Some functions or traits of a wild plant (or animal) is unnecessary for the survival. As a result, many genes of a wild plant are lost in a domestic plant. Wild plants also have more isoenzymes as compared to domestic plants (12). For many fruit tree, such as apple trees, which mainly reproduces via grafting, the genes involved in sexual reproduction are lost and sexual reproduction becomes difficult.

The wild corn species will not survive in wild condition if it has biological traits of a domestic plant, such as a corncob.

In agriculture, people pay much attention to the yield of the plant. The hybrid vigor or quality of a domestic plant almost equates to the concept of high yielding. Thus, the concept of hybrid vigor is meaningful for a specific trait according to our measurements of value. It is not validated in all fields of biology. Moreover, some traits are considered as beneficial for domestic plants, but not for wild plants or animals.

All in all, we conclude that the physiological activities of the wild plants and animals are controlled tightly, which endows its ability to regulate many signaling activities (or functioning) orderly and efficiently.

6. Different Fashions of Biological Regulation of the Domestic Organism

The key idea behind our model is that the nature of biological regulation in the domestic organisms, as compared with wild organisms, shows some differences. The molecular mechanism behind the biological regulation of most biological processes is not clear now and cannot be demonstrated directly and experimentally, thus this conclusion will be confirmed indirectly by the pre-existing facts of biology.

Color patterns of flower are biological traits and often result numerous strains of domestic plants as seen in the example of an orchid (Cymbidium spp.). Flowers can show different color patterns which are changeable in biological evolution.

The color of a flower is determined by many genes as well as regulation fashion of plant physiology. In order to simplify the question, we will not discuss the detailed genes related to color formation in this book. For a given flower or a species which shown mottled patterns, such as red and white mottled pattern, the color pattern of flower is completely controlled by the biological regulation fashion which is formed during the biological development of the flower. Thus, the alteration of the biological regulation fashion can be monitored by a change of color patterns of flower. Some types of orchid flower are shown in Figure 4.

7. Other Facts that Support our Hypothesis

Our hypothesis can explain more observable phenomena and experimental results. Now we will summarize and discuss it briefly as follows.

1. New protein complexes or protein recombination have been well studied in field of biochemistry and have been isolated in a hybrid plant.
2. Teosinte, the ancestor of corn, is a strict short-day plant, but the corn is not sensitive to the length of the day. It indicates that the system development of corn has been changed in

Figure 4. Several color patterns of orchid flower

In Figure 4, we see that color patterns of an orchid flower can have a dominating red area (Figure 4a), be red excessive (Figure 4b), having a red area at central (Figure 4c), red area in the out-ring (Figure 4d), have a vein dominating pattern (Figure 4e), or have sporadic (Figure 4f). Of course, many color patterns of the flower have not been pictured in Figure 4. These facts indicate that regulation fashion of color formation of flower is changeable in biological evolution. In addition, it is a common phenomena of biology and it can be clearly seen in many types of flower or plants.

biological evolution, with the disappearance (or modification) of the photoperiod of corn development; the biological regulation mechanism of it has been changed (13). This phenomenon is also observed in rice. It has been reported that domestic rice has no clear photoperiod.

3. The efficiency of photosynthesis in wild rice is low while the efficiency of photosynthesis of hybrid rice is high. The basal

metabolic rate of hybrid and domestic plants are high while they are usually low in a wild plant.

4. When domestic and wild rice are hybridized, phenotypes of F1 offspring are the same as that of wild rice. It indicated that the wild has the dominating phenotype.
5. The lifespans of grafted tree have been greatly reduced (14).
6. In the hybrid cells between different species of animal, the chromosome of one species is often lost in the cell culture. This results from the incompatibility between different physical states of the biological species (15).
7. Compared to the seed of wild and domestic rice, the seeds of the hybrid rice have many other characteristics. It includes: 1) sensitivity to water, as the seed of hybrid rice swelled rapidly in water and is sensitive to hypoxia; 2) the germination percentage of the seeds of hybrid rice decrease rapidly when they are stored; 3) the seeds of hybrid rice are sensitive to high temperatures during sprouting; 4) hybrid rice easily catches Hemp disease; 5) hybrid rice is sensitive to many viruses and bacteria; 7) hybrid rice is prone to overgrowth. These facts indicate that hybrid rice do not possess better quality trait except for yielding.
8. Many gene mutations of bacteria have been isolated and are valuable in fermentation engineering. All these gene mutations can greatly alter the regulatable mechanism of signaling transduction pathways (16).

As hybrid vigor and domestic plant shares similar working mechanism, and it is predicted that any trait of hybrid vigor can be also obtained in a domestic plant through serial mutation and selection, but in practice, too many gene alterations have been involved in the production of hybrid vigor, it is difficult to obtain such a strain of genes in a domestic plant using the common method of genetics.

8. The Asexual Hybridization and Plant Grafting

Asexual hybridization and plant grafting represent special methods of hybridization (17). When a scion is grafted into a stock, the two parts of the different plants are fused into a new one. Although gene transfers between scion and stock is possible and may occur in some cases, the increase of the yield or the size of fruit cannot be explained by transfer of genes between scion and stock (18). The reason is because all fruits of a grafting tree usually have a similar size, and the gene transfers among different cells are rare. Normally, small molecules can be freely exchanged between scion and stock, materials exchanged between them can act as the regulator for the physiological activities of the scion and thus changes the yield of the plant.

Exchangeable materials mainly include salt, RNA, lipid, protein and sugar, which all have a low molecular weight. These materials can also affect processes of biological development of the scion, a new phenotype can be developed. The genetic background of the scion has all the essential information required for the development of such a phenotype, but it cannot develop itself normally; in this case, materials of the stock must act as the environmental factors for the development of the new phenotypes of scion. Among these new phenotypes of scion, some can be genetically control and can be transmitted by sexual reproduction, but most of them can be only transmitted by grafting and not via the way of sexual reproduction. An example, within which the trait can be transmitted via sexual reproduction, is given in Ref. 19. Within this mechanism gene transferring is unnecessary.

Several reports have indicated that the gene transferring between the scion and the stock truly occurs (17, 20). It belongs to another mechanism for the change of inheritance: gene recombination at molecular level which may not be observable by traditional methods of genetics. The gene recombination in chromosome has been well documented now.

These different mechanisms for the production of a new phenotype of scion has been summarized in Figure 5.

In Figure 5, we have proposed three mechanisms of inheritance that appeal in plant grafting. These mechanisms can account for all observed phenomena that occur in plant grafting.

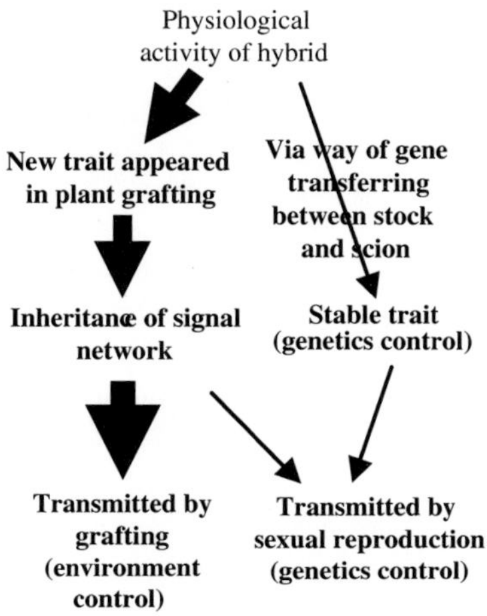

Figure 5. The classification of the inheritance which occurrs in plant grafting

Three mechanisms of inheritance are expressed in this Figure. Most new phenotypes of the scion can be transmitted via asexual reproduction, or grafting, which is the inheritance of signaling network. Some may be transmitted by sexual reproduction. A third fashion of the new phenotype is obtained by gene recombination of the chromosomes and which can be transmitted by sexual reproduction. The thickness of arrow represents the probability of such pathways or mechanisms.

References

1. Darwin CR. The Effects of Cross- and Self-fertilization in the Vegetable Kingdom. John Murry; 1876.
2. Birchler, J. A., Yao, H., Chudalayandi, S., Vaiman, D., & Veitia, R. A. (2010). Heterosis. *The Plant Cell*, 22(7), 2105–2112.
3. Chen, Z. J. (2010). Molecular mechanisms of polyploidy and hybrid vigor. *Trends in Plant Science*, 15(2), 57–71.
4. Liu, Y. S. (2011). Inheritance of acquired characters in animals: A historical overview, further evidence and mechanistic explanations. *Italian Journal of Zoology*, 78(4), 410–417.
5. Hauser, M. T., Aufsatz, W., Jonak, C., & Luschnig, C. (2011). Transgenerational epigenetic inheritance in plants. *Biochimica et Biophysica Acta (BBA)-Gene Regulatory Mechanisms*, 1809(8), 459–468.
6. Chen, Z. J. (2013). Genomic and epigenetic insights into the molecular bases of heterosis. *Nature Reviews Genetics*, 14(7), 471–482.

7. Groszmann, M., Greaves, I. K., Fujimoto, R., Peacock, W. J., & Dennis, E. S. (2013). The role of epigenetics in hybrid vigour. *Trends in Genetics*, 29(12), 684–690.
8. Schnable, P. S., & Springer, N. M. (2013). Progress toward understanding heterosis in crop plants. *Annual Review of Plant Biology*, 64, 71–88.
9. Eggan, K., Akutsu, H., Loring, J., Jackson-Grusby, L., Klemm, M., Rideout, W. M., ... & Jaenisch, R. (2001). Hybrid vigor, fetal overgrowth, and viability of mice derived by nuclear cloning and tetraploid embryo complementation. *Proceedings of the National Academy of Sciences*, 98(11), 6209–6214.
10. Romagnoli, S., Maddaloni, M., Livini, C., & Motto, M. (1990). Relationship between gene expression and hybrid vigor in primary root tips of young maize (Zea mays L.) plantlets. *Theoretical and Applied Genetics*, 80(6), 769–775.
11. Auger, D. L., Gray, A. D., Ream, T. S., Kato, A., Coe, E. H., & Birchler, J. A. (2005). Nonadditive gene expression in diploid and triploid hybrids of maize. *Genetics*, 169(1), 389–397.
12. Brown, A. H. D. (1978). Isozymes, plant population genetic structure and genetic conservation. *Theoretical and Applied Genetics*, 52(4), 145–157.
13. Beadle, G. (1980). The ancestry of com. Scientific Amo'ican, 242, 112–119.
14. Feucht, W. (1987, September). Graft incompatibility of tree crops: an overview of the present scientific status. In *International Symposium on Vegetative Propagation of Woody Species* 227 (pp. 33–42).
15. Croce, C. M. (1976). Loss of mouse chromosomes in somatic cell hybrids between HT-1080 human fibrosarcoma cells and mouse peritioneal macrophages. *Proceedings of the National Academy of Sciences*, 73(9), 3248–3252.
16. Madigan, M. T., Martinko, J. M., Parker, J., & Brock, T. D. (1997). Biology of microorganisms (Vol. 985). Upper Saddle River, NJ: prentice hall.
17. Hartmann, H. T., & Kester, D. E. (1968). Plant propagation: principles and practices. Plant propagation: principles and practices., (Edn 2).
18. Stegemann, S., & Bock, R. (2009). Exchange of genetic material between cells in plant tissue grafts. *Science*, 324(5927), 649–651.
19. Zhang, D, Meng, Z, Xiao, W. *et al.* Graft-induced inheritable variation in mungbean and its application in mungbean breeding. *Acta Botanica Sinica*, 2002, 44(7): 832–837.
20. Bock, R. (2010). The give-and-take of DNA: horizontal gene transfer in plants. *Trends in Plant Science*, 15(1), 11–22.

Chapter 28

Biological Evolution

Summary: From the perspective of irreversible thermodynamics, biological evolution is a specific example for the origin of natural order and is an irreversible process. The principle of biological evolution is discussed in this chapter on the basis of irreversible thermodynamics. The essential idea is that any biological species represents a specific signaling network and the behavior of biological evolution can be described and analyzed by applying the principle of system biology.

As a matter of fact, the biological evolution has been well accepted over the whole scientific community, but we still lack a satisfactory theory which can account for diverse phenomena which we have observed with regard to biological evolution. Thus, the debate about its mechanism has been repeatedly brought up since it was first proposed by Larmak, and as of today, no consensus about it has been obtained (1–7).

Two difficulties regarding the theory of science hinder us in gaining a comprehensive understanding of biological evolution. The first is the problem of the origin of natural order (5–7). As biological evolution is a process typical of the origin of natural order, the traditional thermodynamics (as well as science) is no able to account for such a problem. The second is the system theory. Biological organisms or a cell is a complex system of physiology which can be only analyzed by application of system logic (8). However, traditional physics is established on the basis of the element logic, rather than a system logic itself and thus a purely theoretical explanation about biological evolution cannot be obtained. Fortuitously, these two questions have been theoretically resolved and this makes it possible to formulate an axiomatic theory of evolution theoretically (8–10).

In this chapter, we will discuss the basic principles of the theory of biological evolution on the basis of the bio-signal network theory and irreversible thermodynamics.

1. The Basic Principles

According to the bio-signal theory of molecular biology, any biological organ, at molecular and general levels, represents one type of bio-signal network (or signaling network, a system of physiology) and shows a hierarchical structure (11–13). Biological evolution is a typical process of the origin of natural order, and new species of life are generated in this process.

According to irreversible thermodynamics and system biology, we first proposed the following principles which can be utilized to analyze the features of evolution in following discussions:

Principle 1. **Any biological function (organ, trait, and species) corresponds to one special biosignal network (signal transduction network), the emergence of the new functions or biological organs is called biological evolution**.

Principle 2. **The biological evolution is an irreversible process from the perspective of thermodynamics.** When a new species is produced, it will remain stable for a long time.

Principle 3. **Giant independent spaces (IS) on the basis of any signal network, can be generated; new species of signaling network, can be generated from this giant space.** This enables the generation of numerous species from same type of signaling network. For example, numerous species of bird could be generated from the basis of a wing.

Principle 4. **The bio-signal network shows hierarchical structure. The relationships among different networks, whether at different or the same level of biological organs, vary from species to species.** In order to obtain high viability or survival ability, the signaling activity of these signaling networks should be matched up well. This enables a species to evolve and adapt its living condition well. For example, any species has its particular sequence for a protein.

Principle 5. **The stability of a bio-signal network corresponds to the viability of a species.** Thus, at molecular level, the viability of life arises from the incompatibility of signaling activity, such as lacking essential binding ability between receptor and ligand.

Principle 6. **For any species, there is a specific bio-signal network which controls its evolution and a fundamental bio-signal network on which the species is developed.** For example, the human intelligence is our predominant ability (or functioning); our senses (olfactory receptor) have degenerated in evolution because they are harmful to our thinking ability and other abilities (14). The fundamental bio-signal network for the origin of humans is the nervous system.

2. The Gene Mutation and Biological Evolution

According to traditional genetics, the gene mutation produces variations and selections of favorable variations that results in biological evolution (16, 17). Logically, it can account for all natures of biological evolution, particular to observations obtained from general biology. But the mechanism of biological evolution cannot be analyzed based on this theory. So, we need to find new theoretical foundations that can support our research about biological evolution.

In addition, we believe that the gene mutation is not the dominating factor for biological evolution or speciation and this can be clearly seen in following facts.

1) The mutation rate of different genes (proteins) varies and it implies that it is not the controlling step of biological evolution. The mutation rate of a regulating gene is much high than that of common gene (18). It indicates that the gene involved in biological speciation has high mutation rate.

2) Some types of genes, such as IgG, receptor and olfactory receptors of animal, emerged suddenly in biological evolution.
3) Many olfactory receptor genes of human are pseudo-genes (14). This implies that the evolution of human is accompanied by the degeneration of the olfactory receptors gene. One view held by many people is that the dysfunction of such a gene is helpful to the development of the human thinking ability.
4) The genetic drift concept has been proposed, which indicates that some types of gene mutation have nothing or little to do with evolution (19). This view may not be valid in all cases, but is true for most cases.
5) The Alu sequence of human, which is a highlighted marker for a species, can be produced rapidly and propagated into other sites of all chromosomes in a short time. From the perspective of evolution, the chromosome shows a dynamic structure and changes rapidly (20).
6) The co-evolution phenomenon indicates that the evolution rate of protein sequences is controlled by its function (21).

So we conclude that gene mutation is a fundamental factor, but not controlling factor for biological evolution (22–25). The controlling factor of biological evolution is the formation of new biological function (organ) which can give individual a new ability to overcome the unfavorable conditions or improve the survival skills of individuals.

3. The Driving Force of Biological Evolution

The current view of biological evolution comes from Darwin's theory of evolution and genetics. It states that gene mutation and biological selection is the mechanism for biological evolution. The main idea behind this theory is the "survival of the fittest". However, it gives no explanation to survival ability of life.

The survival ability of a lion living in Africa comes from the perfect functioning of its body, such as its strength of fighting, strength of biting, running speed, visual power and hearing ability. Thus the perfection of biological functioning is the source for its survival ability and biological evolution. Perfect functioning of biological organs depends on its fine structure and the cooperation of its components. Thus, biological evolution is the improvement process of biological functioning and is an intrinsic nature of life. Although a change of environment may play great part in the speciation of some species, it is not the controlling factor which directs biological evolution, such as the origin of the brain.

From the perspective of the signaling network theory, a species cannot guarantee perfect functioning of all types of signaling proteins or organs, and this nature is called incompatibility among different signaling activities. At a molecular level, the sterility of a hybrid results from incompatibility of different signaling activities between their parents. A species can evolve in many directions and anyone of them can guarantee perfect functioning of one or a few types of signaling activity. For example, humans have explored more thinking ability of the brain while wolves have explored their ability to fight more.

In this view, biological evolution can be reconsidered as the evolution of biological functioning. It includes the generation of new functions and the improvement of existing biological functions, as well as the refinement of cooperation efficiency of many different functions (signaling activities). Thus, the driving force of biological evolution is the imperfection in biological functioning.

4. A Hypothesized Mechanism for Biological Evolution

From the perspective of the origin of natural order or developmental biology, a giant space of possible biological functions can be produced from the same genetic foundation or genome. Two important conclusions, which are of importance in biological evolution, can be deduced: 1) some biological functions generated based on the genetics foundation of a body may not be needed; 2) imperfections of some new functions (or promiscuity of normal function) may appear under some cases, such as in extreme condition; in protein science this feature is termed as the promiscuity of proteins.

It has been reported that an English man had gained the ability to utilize echolocation mechanism for navigating after he lost his sight shortly birth (26). As he had lost his visual ability, other possible ability which were normally suppressed, were explored for his survival. This fact indicates that people has much more potential ability than the normal abilities we possess today. It has been hypothesized that bats gained such an ability to navigate via echolocation mechanism in history through the loss of sight in some degree. Dogs, pigs and bird have much better memory as compared to human, albeit their low intelligence. All these facts indicate that brain of an animal has many more abilities than what we have explored or recognized. When these types of potential abilities are utilized under some cases, new species may be generated.

The mechanism of biological evolution is expressed in Figure 1.

Several processes of biological speciation were shown in Figure 1. Initially, a new and unstable function emerged and it favored the survival of the body. Second, the population of these types of individual was increased. Third, the biological function was further

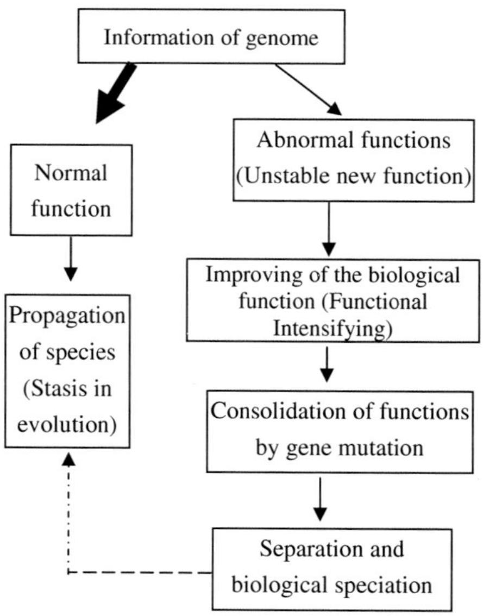

Figure 1. A hypothesized mechanism for biological evolution (speciation).

improved with repeated use and at this stage the function is under environmental control. Fourth, through gene mutation, the desired function was consolidated and become genetically control. Fifth, the new species was produced due to its separation from other normal individuals. The acquired inheritance takes place at steps 3 and 4. Our opinion is that Larmark's theory (or acquired inheritance) can be reconciled with Darwin's theory (gene mutation and natural selection).

5. The Punctuation Equilibrium

When we study the textbook of biological evolution, the first piece of knowledge acquire is that numerous species are developed from the basis of a particular organ. For example, numerous mammals are developed from skeletons, numerous fishes are developed from the gills, numerous insects and birds are developed from wings, numerous species of phylum Mollusca are developed from the shell, birds and mammals are developed from the bones (antisternum). The emergence of such structures or biological organs is a major event in biological evolution. When it appears, the biological evolution, or speciation, will process rapidly. Thus comes the concept of punctuation equilibrium (27).

For the life at the stage of stasis, its genome remains stable at level of the chromosome and their behaviors can be precisely described by laws of genetics. But when a species is subjected to change, the information in the genome is changeable. There are two types of information change of a genome (or gene mutation): one is the giant change of genetics information which includes the duplex of chromosomes, the decrease and increase in chromosomes; another is the fine change of DNA sequence or point mutation of a gene, such as which occurs in protein evolution, and plays an important role in the improvement of protein function under a specific environment.

The problem is that the new function, such as an immune function and olfactory sensation, requires the production of new proteins, like IgG, IgM, olfactory receptors, organization of its component within a particular structure, and the proper biological regulation fashion of this new function. These need dramatic changes of the genome. It is a well-known fact that animal has many copies of some types of DNA sequences, such as the Alu sequence for humans. It was generated at a certain time and rapidly spreads over all chromosomes (28).

Facts indicate that the generation of genes appears first, and then the products of these genes, or protein, can be organized and improved in future. A hypothesized mechanism is expressed in Figure 2.

The origin of triticum aestivum (Wheat) serves as a good example for this mechanism. Chromosomes of triticum aestivum come from Triticum dicoccoides and Aegilops tauschii by accidental hybridization between them (28, 29). This species was further developed into many breeds after it was cultured over the world. The evolution of Triticum dicoccoides, Aegilops tauschii, and triticum aestivum is normal, and the origin of triticum aestivum occured by chance. Both

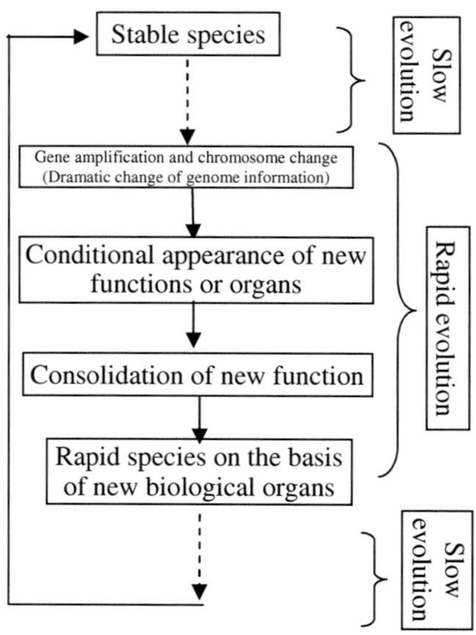

Figure 2. The hypothesized mechanism for the punctuation equilibrium.

The original species have undergone rapid and dramatic gene mutations and information in the genome was greatly increased. Then the new function, such as the wings of birds, was produced. The rapid evolution of the species then occurs and many new species can be produced.

Triticum dicoccoides and Aegilops tauschii are not the ancestor of triticum aestivum.

The Cambrian Explosion is an example for the punctuation theory of biological evolution (30–31). My opinion is that the Cambrian Explosion was induced by the emergence of a primary nervous system and which resulted in the flourish of speciation of lower animals. When the primary nervous system had emerged, had gained a higher survival ability, and its population grow up rapidly. But its components, structures, and properties had not been fixed, or in other words, they are flexible and can be developed in many directions, so new species can be developed in different ways, and this results in rapid speciation.

In history, the origin of the neural system, lung, and backbone, and the major change in these organs, have changed the direction of biological evolution and where many species originated. But it needs a dramatic change of genetic foundation or chromosomes, the underlying mechanism of such a change is still not clear now.

Our theory provided reasonable explanation to the debate between Gradualism and Punctuated Equilibrium. The generation of the new function (or biological organs) must be explained by punctuated equilibrium theory. The change of parameters of a function, such as the size of a bone or wing, should be explained by gradualism theory for biological evolution.

6. The Origin of Humans

We have summarized basic facts about this topic and the following conclusions can be obtained.

First, although humans have a much higher intelligence and physical capacity as other mammals, no basic ability can be utilized to differentiate humans from mammals. Almost all mammals have ability to remember, to think in some degree, to make and utilize tools, albeit a weaker or smaller ability.

Second, humans have a lot of abilities that other species of mammals have. Humans have all families of immune proteins, include of the IgM, IgG, IgD, IgA and IgE. The human eye can see color, while for the most mammals, the world is seen through bright and dark figures.

Third, there is no human gene which can be utilized to differentiate humans from other mammals. The sequence of human and Pan Troglodytes is now known, although many differences in genome have been discovered, no gene can be subtitled for the human gene (32). Thus, at level of molecular biology, they are same. The major difference, which cannot be expressed in DNA sequences, is the fine regulation mechanism of biological functioning or the fine structure of biological organs, which is theoretical foundation for our discussion.

Any type of signaling activity of Pan Troglodytes may differ slightly from a human's, but the cooperation of many signaling activities (which corresponds to a biological function or our ability) differs largely between humans and Pan Troglodytes. So we theorize that humans came about from the process to improve our abilities, such as speaking, thinking. The key difference between humans and Pan Troglodytes is in thinking ability.

The theory provides two possible mechanisms for the appearance of humans. The first may be called the differentiation mechanism. It said that the ancient monkey (or Apo) has all the genetical information of a human, but some undesired genetical information (or suppression gene) hinders a transformation. This undesired information of genomes was lost in evolution via way of genetic drafting. Once the ancient human had appeared, its functioning was further improved by gene mutation.

The second mechanism is that some gene mutations played a key role for the origin of humans and other gene mutations can further improve the human ability by regulating the relation of different functions.

It is possible that the olfactory receptor genes of a human are harmful to the thinking ability of human, thus they have mutated into pseudo-genes in the process of the speciation of humans (14, 15). A possible explanation is that activities of olfactory receptors consume a large amount of energy which is required for the thinking ability of humans.

References

1. Lamarck, J. B. (1809) Zoological philosophy. published by Macmillom, New York, Translated by H. Elliot (1914).
2. Darwin, C. (1859) On the origin of the species by means of natural selection or the preservation of favoured races in the struggle for the life. John Murray, London.
3. Strickberger M. W. (2000) Evolution 3rd Ed. Jones and Bartlett Publishers. Sudbury.
4. Jablonka E. Lamb M.J. (1995) Epigenetic inheritance and evolution: The Lamarckian dimension. Oxford University, Oxford.
5. White, M. C., Marin, D. B., Brazeal, D. V., & Friedman, W. H. (1997). The evolution of organizations: Suggestions from complexity theory about the interplay between natural selection and adaptation. *Human Relations*, 50(11), 1383–1401.
6. Kauffman, S. A. (1993) The origins of order: Self organization and selection in evolution. Oxford University Press.
7. Popp, F. A., & Schommers, W. (2002). What is life?: Scientific approaches and philosophical positions (Vol. 4). World Scientific.
8. Zhao, Q. Nonalgorithmicity and algorithmicity of protein science. *Advance in bioscience and biotechenology*, 2011, 2, 340–346.
9. Zhao, Q. (2001) Irreversible thermodynamics theory for protein folding and protein thermodynamics structure. *Prog Biochem Biophys* 28, 429–435.
10. Zhao, Q. (2013) Nature of Protein Dynamics and Thermodynamics. *Rev. Theor. Sci.* 1, 83–101.
11. Zhao, Q. (2013). A molecular and biophysical model of the bio-signal. *Quantum Matter*, 2(1), 9–16.
12. Zhao Q. (1996) The physical movement of macromolecules and the physical process of signal transduction. *Chinese J of Rehabilitation Theory & Practice* 2, 135–140
13. Ravasz, E., Somera, A. L., Mongru, D. A, Oltvai Z. N. & Barabási A. L. (2002) Hierarchical Organization of Modularity in Metabolic Networks. *Science* 297, 1551–1555
14. Gilad, Y., Wiebe, V., Przeworski, M., Lancet, D., & Paabo, S. (2004). Loss of olfactory receptor genes coincides with the acquisition of full trichromatic vision in primates. PLoS biology, 2(1), E5–E5.
15. Gilad, Y., Man, O., Pääbo, S., & Lancet, D. (2003). Human specific loss of olfactory receptor genes. *Proceedings of the National Academy of Sciences*, 100(6), 3324–3327.
16. Fairbanks D.J., Anderson W.R. (1999) genetics. Thomson Publishing Company.

17. Morgen T. H. (1926) The theory of the gene. Yale University Press. New Heven
18. Borneman, A. R., Gianoulis, T. A., Zhang, Z. D., Yu, H., Rozowsky, J., Seringhaus, M. R., Wang, L. Y., Gerstein, M. & Snyder, M. Divergence of transcription factor binding sites across related yeast species. *Science*, 2007 317, 815–819.
19. Lande, R. (1976). Natural selection and random genetic drift in phenotypic evolution. Evolution, 314–334.
20. Eichler, E. E., & Sankoff, D. (2003). Structural dynamics of eukaryotic chromosome evolution. *Science*, 301(5634), 793–797.
21. Pazos, F., & Valencia, A. (2008) Protein coevolution, coadaptation and interactions. *The EMBO journal*, 27(20), 2648–2655.
22. Barrier, M., Robichaux, R. H., & Purugganan, M. D. (2001). Accelerated regulatory gene evolution in an adaptive radiation. *Proceedings of the National Academy of Sciences*, 98(18), 10208–10213.
23. Levin, B. R., & Bergstrom, C. T. (2000). Bacteria are different: observations, interpretations, speculations, and opinions about the mechanisms of adaptive evolution in prokaryotes. *Proceedings of the National Academy of Sciences*, 97(13), 6981–6985.
24. Tenaillon, O., Toupance, B., Le Nagard, H., Taddei, F., & Godelle, B. (1999). Mutators, population size, adaptive landscape and the adaptation of asexual populations of bacteria. *Genetics*, 152(2), 485–493.
25. Nevo, E. (2001). Evolution of genome–phenome diversity under environmental stress. *Proceedings of the National Academy of Sciences*, 98(11), 6233–6240.
26. William Kremer, Human echolocation: Using tongue-clicks to navigate the world. BBC new magazine 12 September 2012, http://www.bbc.com/news/magazine-19524962
27. Gould S.J. Eldredge N. (1993) Punctuated equilibrium comes of age. *Nature*. 366, 223–227.
28. Kindt, T. J., Goldsby, R. A., Osborne, B. A., & Kuby, J. (2007). Kuby immunology. Macmillan.
29. Zohary, D., Hopf, M., & Weiss, E. (2012). Domestication of Plants in the Old World: The origin and spread of domesticated plants in Southwest Asia, Europe, and the Mediterranean Basin. Oxford University Press on Demand.
30. Cavalier-Smith, T. (2006). Cell evolution and Earth history: stasis and revolution. Philosophical Transactions of the Royal Society of London B: *Biological Sciences*, 361(1470), 969–1006.

31. Morris, S. C. (1989). Burgess Shale faunas and the Cambrian explosion. *Science*, 246(4928), 339–346.
32. Sequencing, T. C., & Analysis Consortium. (2005). Initial sequence of the chimpanzee genome and comparison with the human genome. *Nature*, 437(7055), 69–87.

Chapter 29

Scientific Explanation of Traditional Chinese Medical Theory

Summary: Traditional Chinese medical (TCM) theory developed in the ancient times based on our daily experiences and medical practice. It has incorporated these experiences into a unified philosophical theory of biology. In contrast, modern biology developed on the basis of the scientific methods, which, although extremely valuable, are unable to explain our daily experience. A scientific explanation of TCM will be valuable for expanding the scope of medical science and biology. Reconciliation between the scientific approach and TCM is therefore a key task for theoretical biology. Here, we will show that TCM can be explained by applYing principles of biological theory or system biology. The scientific account of TCM provides a comprehensive understanding of biology and medical science.

Traditional Chinese medical theory (TCM) has explored our daily experience of human physiology and medical practice in ancient times and formulated it into a unified philosophical theory, which is remote from the perspective of modern biology and medical science. A scientific interpretation of TCM would prove to be extremely valuable both for biology and medical science. Without that reconciliation, scientists cannot make use of TCM knowledge in modern medicine and biology. However, all scientists who have devoted themselves to this problem have found it deeply puzzling.

The Signaling network (or system biology) theory is expected to be able to answer all physiological questions on the basis of molecular biology (1, 2). In order to explain general and molecular physiology, three basic hypotheses have been introduced into biology (2):

1) One biosignal corresponds to, at least, one type of protein conformational state (or potherse); all other factors (e.g. drug, environmental factors, biological substance etc.) can act on the protein directly or indirectly and then show their biological function *in vivo*. Thus, all physiological changes can be considered biosignal activities.
2) Signal transduction activities are the functional coupling between proteins (or enzymes); signal conduction is the propagation of signal transduction activity.
3) One biological organ (or function) corresponds to one biosignal network; in other words, a biological function is the output of the activity of biosignal network. It represents a sub-system of physiology.

In this paper, we will show that TCM may be explained theoretically in terms of the biosignal theory or system biology. As this chapter focuses on a basic explanatory theory of TCM, we will not consider more detailed experimental data unless they are of general significance for our understanding of TCM.

1. The Scientific Explanation of TCM Concepts

Corresponding concepts of TCM and biosignal theory. According to the biosignal theory, Qi represents the activities of the signal transduction system (4). Bioinformation flow or energy flow accompanies signal conduction activities. Although the Qi concept of TCM comes from ancient Chinese philosophy, the meaning of Qi in TCM differs from that of Chinese philosophy in principle. We briefly summarized biological properties of Qi here.

- Via the Qi concept, all physiological changes can be linked *in vivo*. In ancient times, people knew no detailed mechanisms of physiological change, but found that biomaterials are changeable and transformable *in vivo*, so they hypothesized they transform into each other via Qi. So, all materials have Qi.

- According to Traditional Chinese philosophy, there is only one type of Qi, which links all changes of universe; but according to TCM, there are many types of Qi such as the Qi of acupuncture points, or of viscera.
- According to TCM, tissue shape controls the activities of Qi, and Qi has concrete properties such as taste and smell. On the other hand, Qi has no shape or any measurable properties according to Ancient Chinese philosophy. The philosophical meaning of Qi is identical to the concept of nothing in western philosophy.

The concept of Qi in TCM can be described by the biosignal theory. We will show that this approach can explain all the properties and features of Qi described by TCM (5).

Other concepts in TCM can be developed on the basis of biosignal and the TCM theory. The logical correspondences between the biosignal theory and TCM are summarized in Table 1.

Yin and Yang concepts. Originally, the Yin and Yang concept was used to describe opposing aspects of matter, such as bright and

Table 1. Corresponding concepts between TCM and the biosignal theory.

TCM Concept	Biological concept and explanation
Qi	Signaling activity (Linking all physiological changes *in vivo*)
Wind	Signal conduction (wind is propagation of Qi). Signal conduction is propagation of signal transduction activities.
Tiny substances	Protein, DNA, medicine, etc. (carrier of Qi and biological function)
Phlegm	Protein complex (Causer of diseases)
Channel	Pathway of signal conduction (Nerve fiber and others)
Shape	Biological structure (proteins, biological organs, etc.)
Shape-Qi	Relation of structure–function of protein (shape is related to Qi)
Yang	Factors that enhance signaling activity (Positive regulation of biosignal)
Yin	Factors that slow down signaling activity (negative regulation biosignal)
Acupoint	Places where many types of biosignals interact with each other efficiently

dark, active and inactive. In the field of developing systems, Yin and Yang also refer to fundamental and advanced matters. For example, neural signals belong to Yang and cell biosignals belong to Yin. The neural signal is a special cell signal that regulates the activities of cell and molecular biosignals.

The following criteria may be used to judge the properties of the signal transduction system:

- Signal transduction activities of short duration and smaller activated areas belong to Yang; those of longer duration and larger activated areas belong to Yin.
- The positive regulators of signal transduction belong to Yang. The negative regulators belong to Yin.
- Regulatory (advanced) signal transduction belongs to Yang; fundamental biosignal activities belong to Yin. The fundamental biosignal (basic metabolism) belongs to Yin. The Yang biosignal regulates the activities of the Yin biosignal.
- A medicine or other factor that increases or strengthens signal transduction activities belongs to Yang; one that decreases or weakens those activities belongs to Yin.

We classify some Yang and Yin properties of signal transduction in Table 2.

We must emphasize that classification of a factor as Yin and Yang depends on the method by which we measure it. We can classify more pairs of Yin and Yang for an object or function when we measure them by different methods. There is no logical coherence between different classifications. For example, compared to adrenalin, insulin belongs to Yin. But compared to collagen, it belongs to Yang.

2. Analyzing the Physiological State of a Biological Organ by Yang and Yin Concept

Although molecular reactions are under tight control *in vivo*, the biosignal network (or biological organs) shows stability (3, 4, 6). In other words, the property or contents of the biosignal network is changeable to some degree without a change of its inner structure.

Table 2. The classification of properties of biological factors by Yang and Yin concept.

	Biosignal	(Mechanism or criteria to judge it)
Yang	High Temperature	Increasing protein flexibility
Yin	Low T	Lowering protein flexibility
Yang	Smaller organic molecules[2]	Enhancing protein flexibility
Yin	Larger organic molecules[3]	Lowering protein flexibility
Yang	cAMP[4]	Increasing cellular metabolism
Yin	cGMP[4]	Increasing cell growth
Yang	Neural biosignal	Shorter duration
Yin	Cellular biosignal	Long duration
Yang	Hormone, growth factors	Regulating biosignal
Yin	Glucose, cell skeleton, O_2	Fundamental metabolism
Yang	Agonist of receptor (or protein)	Effect on receptor activity
Yin	Antagonist of receptor (or protein)	Effect on receptor activity
Yang	Capsaicin[5]	Feeling hot, excitive to cellular activity
Yin	Menthol[5]	Feeling cold, suppressing signaling activity
Yang	agonist	Moves fast
Yin	receptor	Moves slowly
Yang	Growth factors	Excitant to cellular signaling
Yin	Extracellular matrix	Depressing cellular signaling
Yang	Brain, Large intestine, Eye	Outside
Yin	Heart, Lung, liver	Inside

Notice: 1) Temperature is most typical factor among many factors that can influence protein conformational change. It could increase the rate of protein conformational change of all proteins; it can also act as a special stimulus for specific receptors (or protein). 2) Smaller organic substances, such as ethanol and volatile general anesthetics, can increase the flexibility of a protein (receptor) and cell membrane and then increase the rate of signal transduction. 3) Larger organic substances can reverse the effect of smaller organic substances and decrease the rate of signal transduction. 4) Often, the signal transduction by way of cAMP, such as epinephrine, can increase the level of glucose in the blood; the signal transduction by way of cGMP, such as insulin, can decrease the level of glucose in the blood. 5) Capsaicin, which come from chili peppers, can raise the level of metabolism of biological organs while Menthol can lower the level of metabolism. 6) Normally, we do not consider basic metabolism as biosignal activity; but in a broader concept, they are biosignal activities.

We can classify regulative biosignal (Yang) and fundamental biosignal (Yin) activities within any physiological system. Illness is caused by an imbalance of Yin and Yang activities, which may be further caused by many factors such as viral or bacterial infection or gene mutation. Thus, the physiological (or pathological) state of a biological organ can be analyzed by the ratio of regulating biosignal (Yang) to the fundamental biosignal (Yin).

In one way, we can classify the physiological states of a biological organ into nine categories according to the ratio of Yang and Yin biosignal activities.

- If the Yang and Yin activities are both normal, the system is in a good and balanced state.
- If the activities of Yang are very strong and Yin activities are normal, the system is in a state of vigorous Yang.
- If the activities of Yang are very weak and Yin activities are normal, the system is in a state of lacking Yang.
- If the activities of Yang are very strong and Yin activities are weak, the system will be very unstable. Disease will occur. TCM terms this the state of Fire.
- If the activities of Yang are very weak and Yin activities are weak, the system is in a state of inactivity.
- If the activities of Yin are very strong and Yang activities are normal, the system is in a state of vigorous Yin.
- If the activities of Yin are very weak and the Yang activities are normal, the system is in a state of lacking Yin.
- If the activities of Yin are very strong and Yang activities are also strong, the system is unstable.
- If the activities of Yin are very strong and Yang activities are weak, the system is inactive. Disease will occur.

Some of these states are unstable, so the physiological state of the body is changeable.

Note: One biological organism has many different signaling pathways, and many pairs of Yang and Yin can be theoretically formulated. However, this analysis can only describe the physiological state of a biological organ on the basis of just one pair of Yang and

Yin. Similarly, we cannot judge the physiological state of a biological organ using just one criterion, such as cAMP/cGMP (7).

Beside the Yang and Yin concept, other terms to describe the physiological state of a biological organ are Fire, Hot, Warm, Chill and Cold. If one has caught Fire, physiological changes will accelerate; the body temperature will become very high at same time and he will feel hot.

3. Physiological Relationship Between Biological Organs

Although the physiological states of biological organs are to some degree independent, they are nevertheless tightly interconnected. In order to analyze the physiological relationships among biological organs, one basic hypothesis is introduced: **Each biological organ has its particular biosignal transduction system, and pathological symptoms related to these biosignal transduction activities will show predominantly in that organ.** This statement does not imply that other biological organs lack these signal transduction activities, but are only weak.

The physiological states of organs are inter-regulated. When physiological change occurs in one biological organ, it also occurs in others, but it is slighter. Hence, when one organ becomes diseased, some clues about the physiological (or pathological) change may be manifested in other organs located in external parts of the body and this enables us to analyze the physiological change in the internal biological organ.

We summarize the representative signal transduction activities of biological organs in Table 3.

We will now briefly discuss some physiological relationships of biological organs in the following sections, using knowledge of the commonality of TCM and molecular biology here.

Heart, small intestine, brain. Heart function is associated with components of the blood, especially to those related to basic metabolism, such as glucose, amino acids, insulin, other hormones, growth factors and ions. Glucose has a great influence on the integrity

Table 3. The representative signal transduction activities of biological organs.

Yin organ	Yang organ	Outside organ	Typical representative signal activities
Heart	Brain	Smaller Intestine Lingua	Biosignal network with rhythm, Fundamental biosignal or metabolism, (Blood component, amino acid, glucose, salt), Ratio of amino acid
Liver	Gallbladder	Eye Nail	The biosignal activities of bile, the biosignal activities of cell skeleton and matrix, the protein components of blood coagulation (all are structural factors)
Spleen	Stomach	Lip	Protease-like activities, regulating the hemorrhage and coagulation; digesting protein, involving immunity activities
Lung	Large Intestine	Nose	Signal transduction activities of Water-absorbing, receptor of smell, other mechanisms are unclear now
Kidney		Ear	Biosignal activities of androgen, estrogen and corticoid, signal to regulate Ca^{++}, PO_4^{---}, biosignal for the development of bone, pressure sensitive biosignaling

Notice: This table is drawn using common knowledge of biophysics, biochemistry, and molecular biology, as well as TCM. Other methods of classification are possible. But this classification can explain more things with regards to TCM.

of the cell membrane. It binds to many species of protein. It also serves as an essential energy supply for the organism and cells. Some species of amino acids, such as glutamic acid, are neurotransmitters. They profoundly influence the activity of nerve cells. Their ratios play an important role for our spirit as well. Salt is necessary for the initiation and maintenance of biosignal activities. When the balances among these materials are disturbed, neural signals will become disordered and abnormal physiological states may appear. Thus, a patient may feel weak and will lose his spirit when the balances in the components of blood are disturbed. The heart represents fundamental biosignal activities; the brain represents advanced biological functions such as one's mental and neural states.

The pumping of the heart is rhythmic. Blood pulse and neural conduction are affected by the same biophysical mechanism; some physiological changes in the brain are transmitted to the blood and we detect them by an abnormality of the pulse. Ancient Chinese doctors diagnose a disease of the brain by the symptoms manifested in the patents blood pulse.

The small intestine absorbs glucose, amino acids and other small molecules; its activities directly affect the heart's functional stability.

Liver, gallbladder, eye, nails. The liver represents two different biosignal activities. One is the cell matrix (protein), cytoskeleton and proteins involved in blood coagulation; these materials sustain the structure of cell and tissues. The other is bile (fat), which supports cell membrane integrity. These materials regulate the excitability of signal transduction systems and have a great impact on their stability.

Many organic substances with higher molecular mass, such as cholesterol and bilirubin (the key constituent of gall-stones), are produced in the liver. Cholesterol decreases the flexibility of the cell membrane and lowers biosignal transduction activity. Ancient Chinese doctors utilized cholesterol and other molecules produced in the liver to decrease Fire (or unnecessary biosignal activity) in the body. Thus, in view of TCM, the liver is related to the activity of anger (excessive state of signaling activities). This phenomenon is explicable because cholesterol and similar molecules can modulate biosignal transduction activity (neural conduction). However, this topic has been insufficiently researched in scientific terms.

The liver is rich in vitamin A, which is necessary for the activity of the eye (vision). The nail is rich in cell matrices. These organs are tightly related to each other from the viewpoint of physiology.

Stomach and spleen. The functions of the stomach and spleen are all associated with protease activity. Proteases are important biosignals *in vivo* (8). Therefore, the stomach and spleen represent the signaling activities of proteases *in vivo*. According to TCM, an abnormality of the spleen can induce hemorrhage and the coagulation of blood. We now know that proteases modulate the production and dissolution of blood coagula. Some blood proteins are synthesized in the liver, and the spleen regulates the blood state: hemorrhage or coagulation. TCM has concluded that the liver is the source of blood and the spleen controls the state of the blood. Diseases of the liver can influence the physiology of the spleen.

Kidney and ear. The kidney is the biological organ of excretion. It directly affects the stability of the blood. It modulates the blood concentrations of Ca^{++}, Mg^{++}, Na^{+}, K^{+}, HPO_4^{--} and so on.

Ancient Chinese doctors have mentioned the great differences between the male and female body. Men and women have different genitalia, which are close to the kidney. It is well known that many hormones, such as androgens, estrogens and corticoids, are involved in the development of man and women. The kidney regulates the Ca^{++} and HPO_4^{--} concentration in the blood and Ca^{++} and HPO_4^{--} are abundant in bone, thus we can conclude that the functions of the kidney are directly related to bone health.

Although its molecular mechanism is currently unclear, the kidney is sensitive to blood pressure and responds sensitively to it. The ear responds to sound waves by sensing the pressure generated in them. We can hypothesize there are common mechanisms behind these biological functions. According to TCM, the ear behaves as a peephole for the kidney. Modern scientific research reveals that many drugs are harmful to both the ear and the kidney. Some molecular biological studies have confirmed the physiological relationship between kidney and ear (9–11).

Lung, large intestine and nose. As there are very little reports on the physiology of lung, we are unable to discuss their relation experimentally. But we can analyze it on the basis of the essential function of these organs.

The lung is the biological organ absorbing oxygen and has a high blood pressure, however it does not excrete water. Why is that so? The biological structure of the lung could inhibit the concretion of the water from the Lung. The large intestine absorbs water *in vivo*. The common physiological feature of lung and intestine is to maintain the water supply of body. Modern studying has revealed some pathological relationship of the lung and the nose as well (12)

Due to constraints, we will not discuss other physiological relationship of biological organs. But these relationships are understandable on the basis of physiology. According to TCM, the fine structure of cell matrix of liver and lung differs from each other in some aspects, but there is no experimental report about it.

4. Channel and Acupuncture Anaesthetic

According to TCM, Qi is transmitted along channels; the acupuncture point is a place at which different channels met up and many types of Qi interact with each other. According to molecular biology, the channels are the pathways along which biosignal is transmitted. The nerve fiber, blood vessel, lymphatic vessel, and pathway of cell communication are all channels. Thus, the Acupuncture points are sites at where different biosignal activities (neural, tissue, cell, blood, lymph and environmental factors) can interact with each other efficiently (4, 13). Of course, some types of Acupuncture points may not be associated with all these biological organs and different Acupuncture points are associated with different nerve fibers. Cells at large acupuncture point are specialized for their physiology due to the number of biosignal activities it is involved. Acupuncture points are fixed entities that are generated during development of the body (14). Smaller Acupuncture points cannot be diagnosed by common medical methods. When we observe a disease at a particular place in our body, some physiological changes will occur and there will be more signaling activities; the small Acupuncture point in this place will become the predominating Acupuncture point, and we can see or feel it. In the perspective of TCM, all sweat glands and pores are smaller acupuncture points, but they have no value for medical application unless they become predominating Acupuncture points under the condition of disease.

This conclusion is supported by many experimental data. Acupuncture points normally show high electric conductance (14–16), high gap junctions between cells (17), higher temperature (17), higher metabolic rate and carbon dioxide release (18) and other signal transduction activities (19). The bio-signaling activity of an acupuncture point can be transmitted into varied biological organs (20, 21).

Acupuncture anaesthetic. The discovery of acupuncture anaesthetic in 1970's convulsed the whole scientific community. But detailed mechanism remains obscure now. We could analyze it using the biosignal theory. When an acupuncture point is stimulated, many types of signaling activities are initiated. This induce the leakage or the secretion of many materials from cells, such as growth factors and cytokines. Some of these materials may inhibit neural conduction or decrease the excitability of nerve cell, all these can change the properties of neural conduction and may induce anesthesia state of the body. Although we cannot indicate which materials induce the acupuncture anesthesia, we predict there are many materials *in vivo* that can do so. Despite the lack of direct experimental evidence, we believe that Acupuncture anesthesia is produced by different molecular mechanisms (22).

Pathways of channels. According to modern biology, the main signaling conduction pathways are nerve fibers and their positions are known to us now. But as these pathways do not meet up with pathways expressed by the Channels figure, debate has ensued (23). We will discuss it here briefly to account for wrong misunderstanding people may have.

In our skin, a signal conduction pathway (cell communication) along some acupuncture points is formed during the development of our body. The signal can be conducted with velocity about 0.5–200 mm/s (2). But it is only one type of channel; it cannot represent all channels for channels can link up all physiological organs physiologically *in vivo* according to TCM. Our opinion is that main channels are nerve fibers for they can link up all biological organs physiologically.

The opinion, that channels are special mechanisms for Qi conduction and there are special structures of channels, has misled many scientists over the world. It should be clarified now.

Can signaling activities be classified into 12 groups? According to TCM, channels and acupuncture points can be classified into 12 basic groups. No scientific evidence is available now as to why signaling activities of human body could be classified into 12 basic types. But some biological phenomena imply that this classification is reasonable. For example, both brain and spinal cord branch out 12 nerve fibers. It implies that all nerve biosignal may be classified into 12 categories. However, this prediction should be confirmed in the future.

5. Pharmacology of TCM

According to the protein signaling theory, all materials of nature and organisms have ability to affect signaling activity of the human body by diversifying mechanisms under specific cases. Theoretically speaking, they all are medicine.

In ancient times, people were unable to measure diversified biological effects of medicine scientifically, so they have explored our daily sensations induced by medicine in their studying of pharmacology. Within this model, the medical effect of drugs can be classified and analyzed with reference to the physiological effects of temperature. Through this, TCM has classified all drugs into two categories: cold or hot drugs. According to signaling theory of medicines, four basic types of drug can be further categorized (2, 13):

1. Hot drugs: they can increase the rate of protein conformational change unspecifically, and includes high temperature, wine, smaller organic substances, and volatile general anesthetics.
2. Hot drug: they can induce protein thermodynamic structural changes of specific proteins and trigger the signaling activity of a protein, and this biosignal activity can increase the metabolic rate of the body and induce a hot sensation of the body, and includes, chili peppers, excitatory transmitter.
3. Cold drugs: they can decrease the rate of protein conformational change unspecifically, and includes low temperature, cholesterol, and bezoar by various mechanisms.
4. Cold drugs: they can induce protein thermodynamic structural changes of specific proteins and initiate signaling activity of a

protein, and this type of signaling activity can decrease the basic metabolic rate of the body and may induce the cold sense of the body. Such drugs include mint, inhibitory transmitter.

The coherence between TCM and the signaling theory comes from the studying of molecular biology of medicine. High temperature and capsaicin (key component of Chili peppers) exert their physiological effect via the same receptor (24, 25). Low temperature and menthol (key component of mint plant) exert their effect via the same receptor (26, 27). The exertion of their effects via the same protein is not a rule, but these facts indicate that TCM is understandable with the knowledge of modern biology.

Vanilloid TRPs respond to temperature, chili pepper, and osmotic changes (25); the chili pepper has been widely utilized by ancient Chinese people to defend against wild humidity. The *Drosophila* TRP channel involves human vision and mint has been utilized to treat eye diseases in the medical practice of TCM. If TCM is right, we can predict that the TRP receptor (or other receptors of menthol) will have effect to increase the stability of cellular signaling activities. This prediction is supported by facts: 1) the menthol is able to induce anesthesia state of human body (28); 2) menthol is the agonist of the GABA receptor, key receptor to regulate cellular excitability of nerve cells (29); 3) menthol Inhibits 5-HT3 Receptor mediated currents, and also suppresses the excitability of a nerve cell (30). Thus, we conclude that most cold drugs have tendency to suppress the excitability of a cell.

Most herbs used in TCM (Traditional Chinese medicine) are complex intermixtures. The classification for it is done according to their entire effect. Thus, a cold drug may contain a component that is a hot drug, and vice versa.

A drug may exert their biological effects mainly on particular organs, and this feature provides a way to classify drugs according to its target. If one drug affects physiological state of the heart, TCM would classify it a "heart drug". Most of the time, one drug can exert their effect on several biological organs.

In conclusion, TCM is understandable by applying the principle of protein signaling theory. Further studying on the molecular mechanism of TCM can answer all the questions we have of it. On the other hand, the knowledge of TCM provides essential cues to the molecular mechanism of a drug. The combination of TCM and molecular biology is helpful to our study of modern medical science, as well modern biology in many aspects.

References

1. Ravasz, E., Somera, A. L., Mongru, D. A., Oltvai, Z. N. & Barabási, A. L. (2002) Hierarchical Organization of Modularity in Metabolic Networks. *Science*, 297, 1551–1555.
2. Zhao, Q. (2013) A molecular and biophysical model of the biosignal. *Quantum Matter*, 2(1), 9–16.
3. Zhao, Q. (2016) Nature of protein dynamics and thermodynamics. *Reviews in Theoretical Science*, 4(1), 1–9.
4. Zhao, Q. (1996) Physical movements of macromolecules and physical processes of biosignal conduction. *Chinese J. of Rehabilitation Theory and Practice*, 2, 135–140.
5. Yin, H. (1992) Fundamentals of Traditional Chinese Medicine. Beijing: Foreign Language Press.
6. Taussig, R., Ranganathan, R., Ross, E. M. & Gilman, A. G. (2002) Overviews of the Alliance for cellular signaling. *Nature*, 420, 703–706.
7. Goldberg, N. D., Haddox, M. K., Nichol, S. E., Glass, D. B., Sanford, C. H., Kuehl, F. A. & Estenssen, R. (1975) Biological regulation through opposing influences of cyclic GMP and cyclic AMP: the Yin-Yang hypothesis. *Adv Cyclic Nucleotide Res*, 5, 307–330.
8. LeMosy, E. K., Hong, C. C. & Hashimoto, C. (1999) Signal transduction by a protease cascade. *Trends Cell Biol.* 9, 102–7.
9. Johnson, K. R., Cook, S. A., Erway, L. C., Matthews A. N., Sanford L. P., Paradies, N. E. & Friedman, R. A. (1999) Inner ear and kidney anomalies caused by IAP insertion in an intron of the Eya1 gene in a mouse model of BOR syndrome. *Human Molecular Genetics*, 8, 645–653.
10. Chrobok, V., Simakova, E. & Liang, J. N. (2000) Residual mesenchyme in the middle ear cavity of the newborn with malformation of the kidney. *Clin Otolaryngol*, 25, 526–30.

11. Meyer zum Gottesberge, A. M., Balz, V. & Felix, H. (2002) Mpv 3 7 Mouse Strain-A Model for the Relationship between the Kidney and the Inner Ear. 84–90.
12. Togias, A. (1999) Mechanisms of nose-lung interaction. *Allergy* 54, 94–105.
13. Zhao, Q. & Wang, A. (1999) The scientific interpretation of traditional Chinese medical theory. (In: The proceeding of excellent medical papers of Chinese's, military publisher, Beijing) pp. 3235–3246.
14. Bowsher, D. (1998) Mechanisms of acupuncture. In: Filshie, J., White, A., editors. Medical Acupuncture. Edinburgh: Churchill Livingston, pp. 69–82.
15. Comunetti, A., Laage, S., Schiessl, N. & Kistler, A. (1995) Characterisation of human skin conductance at acupuncture points. *Experientia* 51, 328–31.
16. Charles, S. (2000) Mechanism of acupuncture-beyond neurohumoral theory. *Medical Acupuncture Online Journal*, 11, 2.
17. Zheng, J. Y., Fan, J. Y., Zhang, Y. J., Guo, Y. & Xu, T. P. (1996) Further Evidence for the Role of Gap Junctions in Acupoint Information Transfer: Intradermal Injection of EDTA Blocks the Acupuncture Effect of Experimentally Induced Arrhythmia. *Am J Acupunct*, 24(4), 291–296.
18. Eory, A. (1984) *In-vivo* skin respiration (CO2) measurements in the acupuncture loci. *Acupunct Electro-Ther Res*, 9, 217–23.
19. Jang, M. H., Shin, M. C., Lim, B. V., Kim, H. B., Kim, Y. P., Kim, E. H. & Kim, C. J. (2003) Acupuncture increases nitric oxide synthase expression in hippocampus of streptozotocin-induced diabetic rats. *The American Journal of Chinese Medicine*, 31(02), 305–313.
20. Liu, Y. K., Varela, M. & Oswald, R. (1977) The correspondence between some motor points and acupuncture loci. *American Journal of Chinese Medicine*, 3, 347–358.
21. Wong, W. H. & Brayton, D. (1982) The physiology of acupuncture: effects of acupuncture on peripheral circulation. *Am J Acupunct*, 10, 59–63.
22. Ceniceros, S. & Brown, G. R. (1998) Acupuncture: A Review of Its History, Theories, and Indications. *Southern Medical Journal*, 91, 1121–1125.
23. Stux, G. (1997) Basics of Acupuncture. New York: Springer-Verlag.
24. Caterina, M. J., Schumacher, M. A., Tominaga, M., Rosen, T. A., Levine, J. D. & Julius, D. (2002) The capsaicin receptor: a heat-activated ion channel in the pain pathway. *Nature*, 389: 821–24, 1997.
25. Clapham, D. E. (2002) Hot and cold TRP ion channels. *Science*, 295(5563), 2228–2229.

26. Mckemy, D. D., Neuhausser, W. M. & Julius, D. (2002) Identification of a cold receptor reveals a general role for TRP channels in thermosensation. *Nature*, 416, 52–8.
27. Peier, A. M., Moqrich, A. Hergarden, A. C. Reeve, A. J., Andersson, D. A., Story, G. M., Earley, T. J., Dragoni, I., McIntyre, P., Bevan, S. & Patapoutian, A. A. (2002) TRP Channel that Senses Cold Stimuli and Menthol. *Cell.* 108, 705–15.
28. Liu, B., Fan, L., Balakrishna, S., Sui, A., Morris, J. B. & Jordt, S. E. (2013) TRPM8 is the principal mediator of menthol-induced analgesia of acute and inflammatory pain. PAIN®, 154(10), 2169–2177.
29. Zhang, X. B., Jiang, P., Gong, N., Hu, X. L., Fei, D., Xiong, Z. Q. & Xu, T. L. (2008) A-type GABA receptor as a central target of TRPM8 agonist menthol. PLoS One, 3(10), e3386.
30. Ashoor, A., Nordman, J. C., Veltri, D., Yang, K. H. S., Shuba, Y., Al Kury, L. & Oz, M. (2013) Menthol Inhibits 5-HT3 Receptor–Mediated Currents. *Journal of Pharmacology and Experimental Therapeutics*, 347(2), 398–409.

Index

Series on the Foundations of Natural Science and Technology

(Continued from page ii)

Vol. 1 *Space and Time, Matter and Mind: The Relationship between Reality and Space-Time*
by W. Schommers

Forthcoming:

Vol. 10 *Space-Time Reality: Mach's Principle, Relativity and Novel Aspects*
by W. Schommers

Vol. 11 *The Origin of Mind: Interplay between Matter and Mind*
by W. Schommers